NUCLEIC ACID SEQUENCE ANALYSIS

COLUMBIA SERIES IN MOLECULAR BIOLOGY

Ernest Borek, Advisory Editor

Molecules and Evolution, Thomas Hughes Jukes
Virus-Induced Enzymes, Seymour S. Cohen
The Cancer Problem: A Critical Analysis and Modern Synthesis, Armin C. Braun
The Modified Nucleosides in Nucleic Acids, Ross H. Hall
Of Microbes and Life/Les Microbes et la Vie, Jacques Monod and Ernest Borek, editors
Control Mechanisms and Protein Synthesis, S. D. Wainwright
Nucleic Acid Sequence Analysis, Stanley Mandeles
Information Theory and the Living System, Lila L. Gatlin

Nucleic Acid Sequence Analysis

STANLEY MANDELES

COLUMBIA UNIVERSITY PRESS
NEW YORK AND LONDON 1972

Stanley Mandeles is Professor and Chairman of the Department of Chemistry at Douglass College of Rutgers University, New Brunswick, New Jersey

Library of Congress Catalog Card Number: 79-186389
ISBN: 0-231-03130
Printed in the United States of America

To Francine

ACKNOWLEDGMENTS

Any virtue to be found in the following pages is due in large measure to the counsel and encouragement of Professor I. Tinoco, Jr., Department of Chemistry, University of California. I am indebted also to Frances Davis, Roselyn Ferreira, David Garfin, Philip Borer, and Mark Mandeles for their expert technical assistance. I thank the National Institutes of Health for support through Grant No. GM 12158, and the National Aeronautics and Space Administration for support through Grant No. NsG 479.

CONTENTS

NUCLEIC ACID SEQUENCE ANALYSIS

1.

INTRODUCTION

WHY DETERMINE SEQUENCES?

It seems certain that the master plan for living, reproducing cells is encoded in deoxyribonucleic acid (DNA). From its molecules comes the information to sustain the homeostatic, adaptive, and replicative mechanisms that are necessary for life to persist. What form, then, does this information take? The evidence is that the biological information resides in the linear sequence of bases on the polynucleotide backbone in almost perfect analogy with the printed material on this page. First, there is a biological alphabet of four letters. These are symbols for the four bases found in DNA: A = adenine, C = cytosine, G = guanine, T = thymine. Words are composed of any three of the four letters and are translated as specific amino acids. Because there are only sixty-four possible arrangements of three letters from an alphabet of four, the dictionary consists of sixty-four words. A set of these words constitutes a sentence that is an explicit instruction for the assembly of specific amino acids into a distinctive protein. The sentence is also an implicit instruction to perform some functional or structural task. A combination of sentences, or a paragraph, often contains the information required for a cell's successful response to an environmental change. Finally, enough of these paragraphs constitute a book (of life, if you will).

The point of this analogy is not its accuracy. It is the hypothesis that all of life's chemical functions and characteristics are ultimately specified by the sequence of bases in nucleic acids. It should be noted that the role of nucleic acids is not limited to the status of a repository or tape of biological information. Ribonucleic acids (RNA), in particular, participate in the expression of this information. The outlines of this process have become clear. Portions of the DNA are copied to specific messenger RNAs (mRNA) which are then translated into corresponding specific protein structures. Other portions of DNA are copied to form transfer RNA (tRNA), 5S RNA, and ribosomal RNA (rRNA). These molecules play essential roles in the translation of base sequences but do not themselves carry the encoded protein sequence.

In the face of this situation one wonders what there is about mRNA that permits it to be translated into an amino acid sequence while tRNA, 5S RNA, and rRNA serve only as part of the translation mechanism. The answer must be traceable to the differences in the base sequences. The most recent evidence indicates that tRNA, 5S RNA, and rRNA contain sufficient numbers of biochemically modified bases or are so highly structured that they may not be recognizable as mRNA and therefore may not perform as messengers. Furthermore, mRNA may contain specific sequences that signal the initiation of the translation process, and these may be missing from the other varieties of RNA. In any event, it is difficult (if not impossible) to dissociate any property or function of nucleic acids from dependence on base sequence. Knowledge of the base sequence of any nucleic acid is essential for rigorously understanding its function in biological systems.

Recently, answers have been obtained to such fundamental questions as:

1. Can a linear correspondence be demonstrated between the sequence of amino acids in a known protein structure and the sequence of bases in its mRNA? (Yes, first found in the case of coat protein of bacteriophage R17 and a portion of its RNA; Adams *et al.*, 1969.)

2. Are there any base sequences in mRNA that do not code for amino acids? (Apparently so in the R17 RNA initiation areas, Steitz, 1969, and intercistronic areas, Nichols, 1970.)

3. Is the genetic code degenerate? (Yes, at least for bacteriophage RNA; Adams *et al.*, 1969.)

4. In a population of messenger RNA molecules specific for one protein, are there any variations in base sequences? (None determined to date in bacteriophage RNA, one variation in a plant virus.)

5. To what extent do tRNA molecules occur that bear the anticodons for the alternate codons in mRNA? Is the sequence of these other tRNAs the same except for the alteration in the anticodon segment? What are the species differences in the tRNA for any one amino acid? (The data bearing on this last point are rapidly accumulating; see Jukes, 1970.)

6. What are the functions of 5S RNA and rRNA, and to what extent are these functions dependent on specific sequences? (Under study in several laboratories.)

These questions have been directed mainly to RNA structure. Of equal importance and interest are the details of DNA structure. Sequence analysis of DNA should augment the information obtainable from genetic mapping experiments and lead to a better understanding of the actual physical structure of the "taped" genetic instructions. Answers to questions concerning the extent of redundancy, or the mechanism of replication of DNA and the synthesis of RNA copies, undoubtedly depend on sequence information.

Another area of interest is the possibility of the correction of genetic defects through use of available molecular biology techniques. It is plain that detailed information concerning the sequence of bases in DNA will be necessary before any specific action of this sort can be taken. It is hoped that the decisions to utilize the available technology will be made with wisdom in view of the far-reaching effects of such changes.

Use of sequence information in the control of nucleic acid-directed disease processes can be made with a great deal less trepidation. Such diseases include influenza, leukemia, cancer, lupus erythematosis, and some other degenerative types. Knowledge of the base sequences in these errant or intruding nucleic acids should prove important in the prevention, control, and cure of these diseases.

In view of the many compelling reasons, both theoretical and practical, for engaging in nucleic acid sequence analysis, it is unnecessary to invoke the Mount Everest challenge, "because it's there." There is some attraction, however, in such a seemingly insurmountable problem.

The number of laboratories committed to the determination of base sequences in nucleic acids is rapidly increasing. In the case of RNA, this is partly due to the encouraging successes in the determination of the entire primary structure of 15 tRNAs, two 5S RNAs, and some internal and end sequences in viral and ribosomal RNA. These successes were in turn due to the development of a number of laboratory procedures of great sensitivity and reliability. There are many investigators whose interest in nucleic acid sequences is part of some other problem such as protein and nucleic acid synthesis or biochemical evolution. In order to accommodate a range of concerns, this manual will attempt to provide the working details of the various methods that have produced sequence information, as well as description of some potentially useful procedures that have been proposed but not exploited for one or more reasons. It is hoped that the inclusion of the latter procedures will serve as a stimulant to implementation or to further refinement for greater feasibility.

The first portion of the manual presents some basic considerations in nucleic acid sequence analysis, followed by descriptions of the various sequence strategies. In each of these strategies, one or more basic laboratory procedures are deemed important enough to warrant description in greater detail, and they are so indicated and presented in the second portion.

STRUCTURES AND SYMBOLS

In most treatments of chemical structure, some early agreement must be reached on the symbols to be used. For nucleic acids, the conventions recommended by the IUPAC-IUB Commission on Biochemical Nomenclature (1970) will be used to represent structure. As indicated in the names, ribonucleic acid (RNA) and deoxyribonucleic acid (DNA), the distinguishing characteristic is the presence of the sugar ribose in the former and 2-deoxyribose in the latter. The nucleic acids are polymers of the monomer nucleotides, and these consist of a purine or pyrimidine base, a sugar (ribose or deoxyribose), and a phosphate. Figure 1.1 shows the line structural formulas for some of

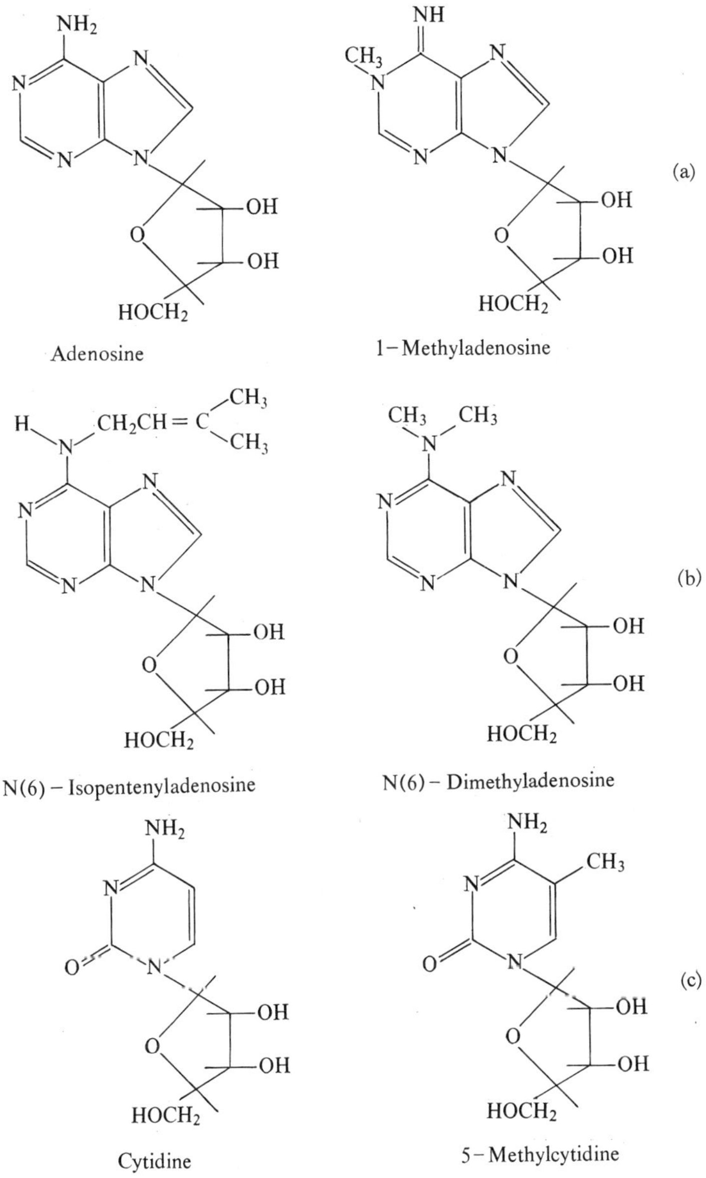

Fig. 1.1. Nucleoside structures.

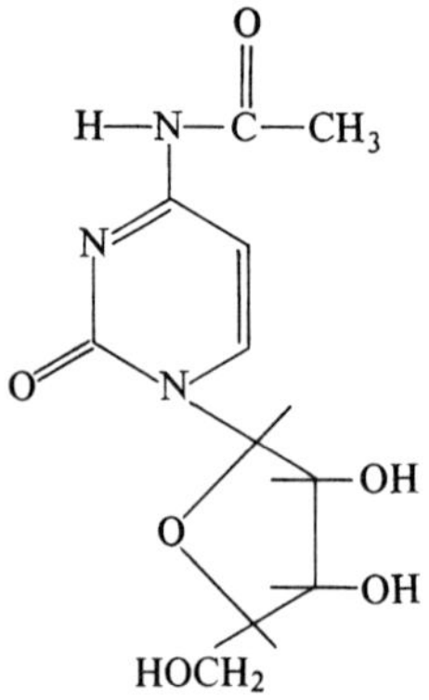

N(4) – Acetylcytidine

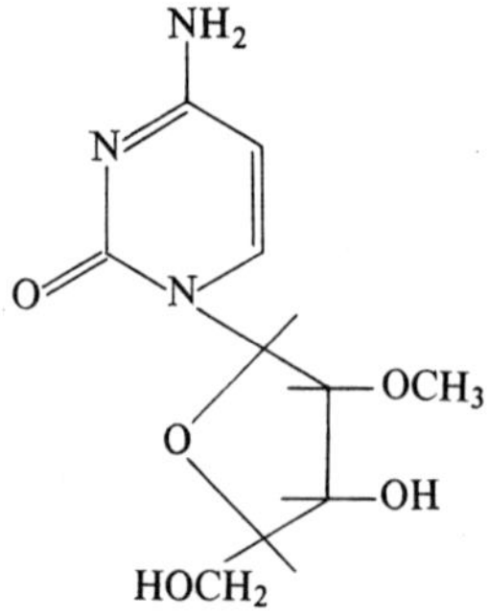

2′–O–Methylcytidine

(d)

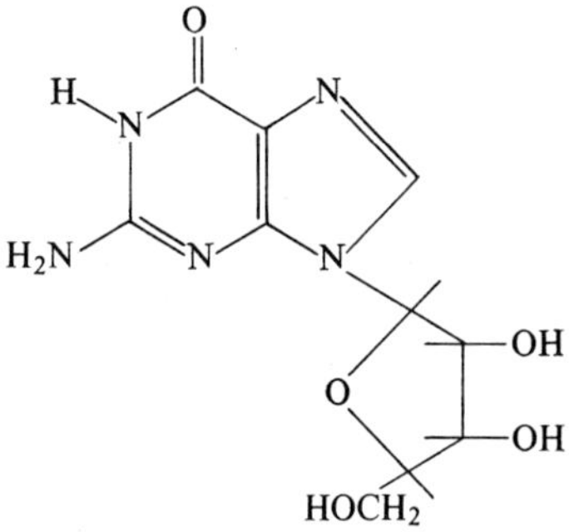

Guanosine

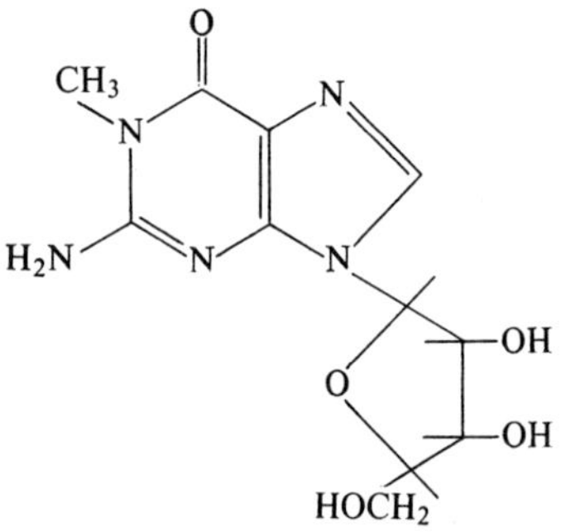

1–Methylguanosine

(e)

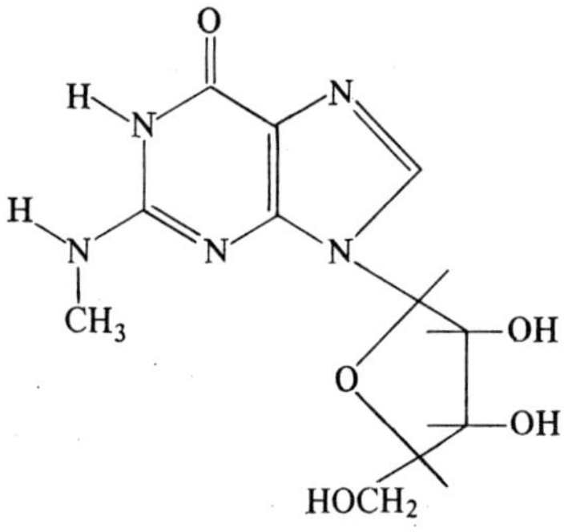

N(2) – Methylguanosine

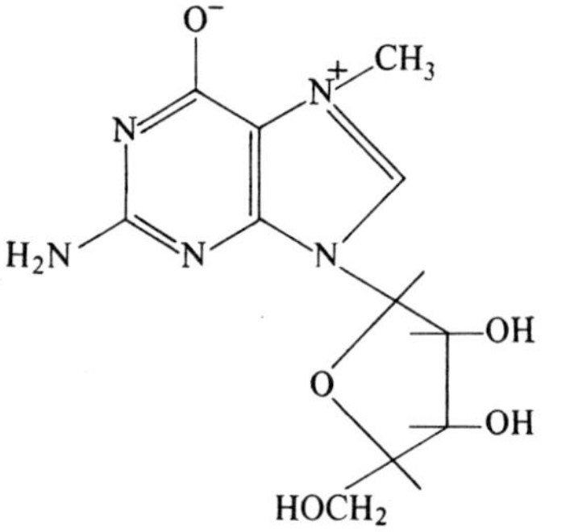

N(7) – Methylguanosine

(f)

Fig. 1.1 (continued)

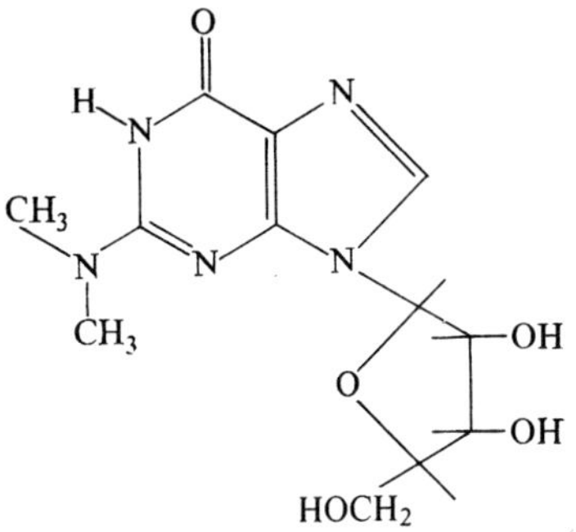

N(2)-Dimethylguanosine

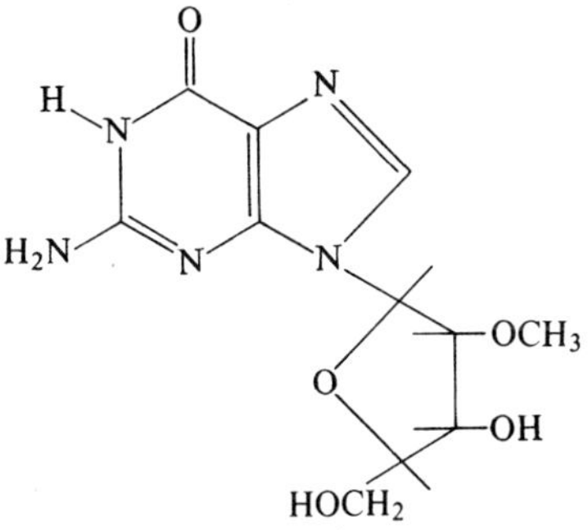

2′-O-Methylguanosine

(g)

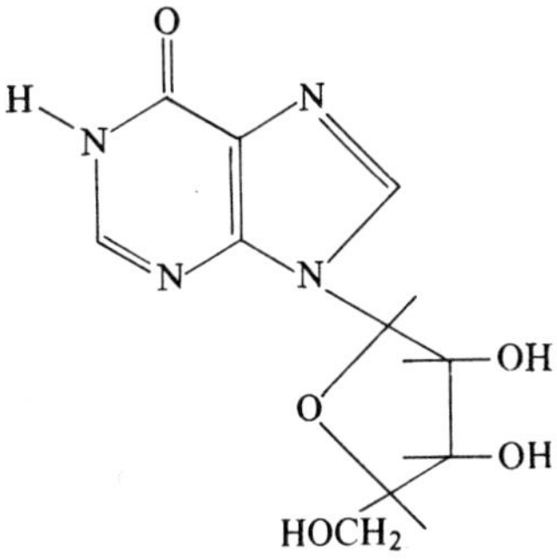

Inosine

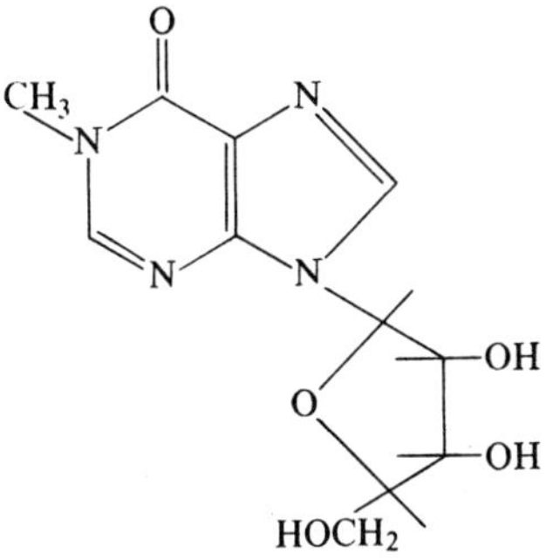

1-Methylinosine

(h)

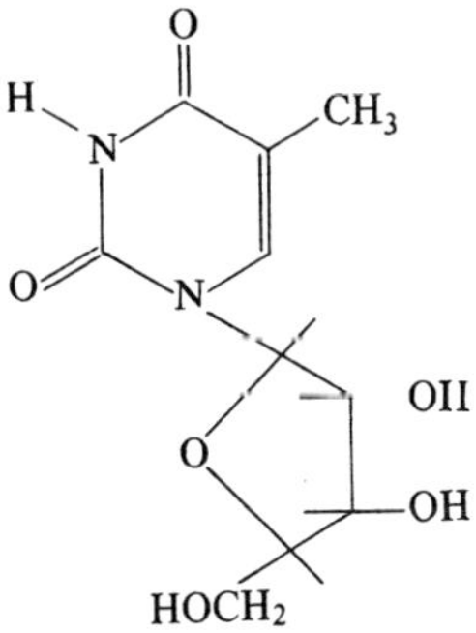

(i)

Ribosylthymine

Fig. 1.1 (continued)

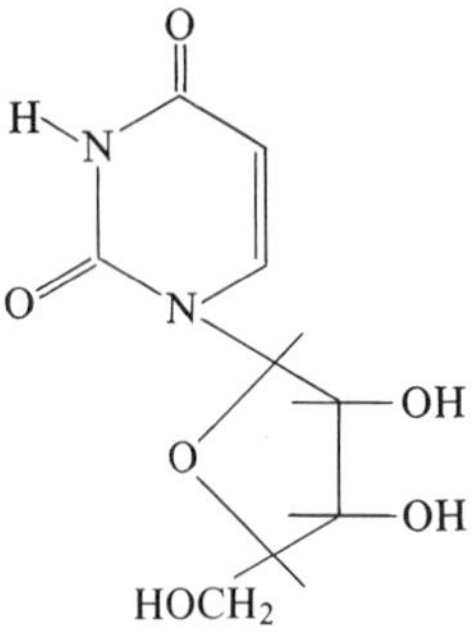

Uridine

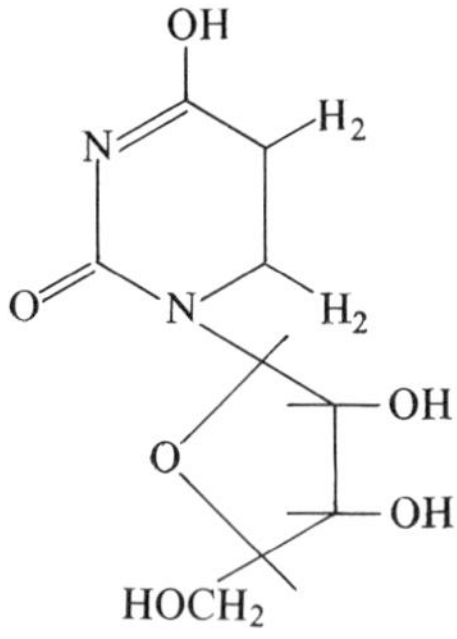

5,6–Dihydrouridine

(j)

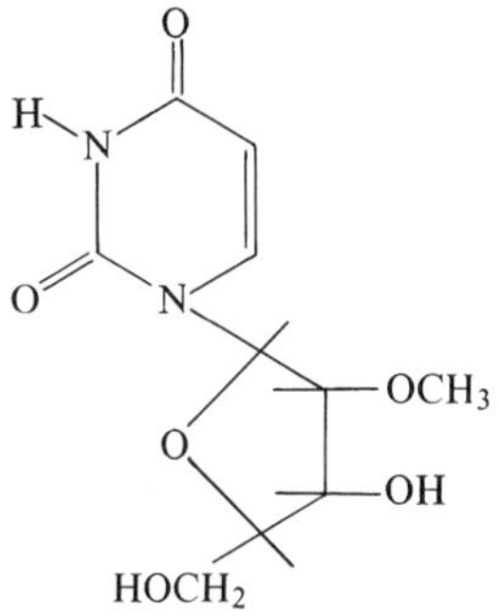

2′–O–Methyluridine

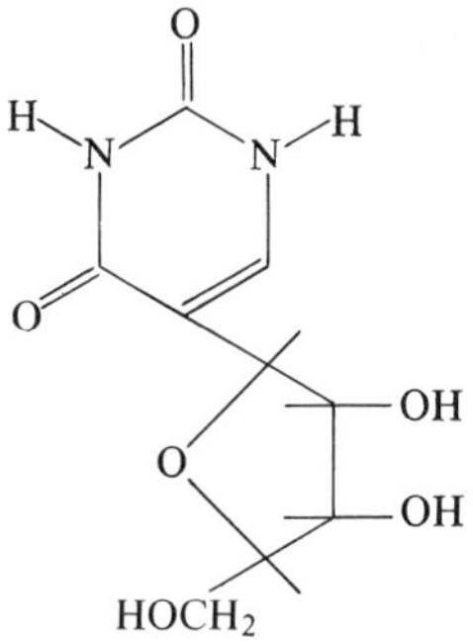

Pseudouridine

(k)

Fig. 1.1 (continued)

the purine and pyrimidine bases found in nucleic acids. They are shown here as ribonucleosides, and they include the four major bases found in RNA (adenine, cytosine, guanine, and uracil) and DNA (adenine, cytosine, guanine and thymine), and the minor bases whose positions in nucleic acids have been determined by sequence analysis. The possibilities for keto-enol tautomerism are purposely ignored in favor of presenting single conventionalized structures.

Abbreviations for these structures that have been developed over the years are listed in column 1 of Table 1.1 (see Dayhoff and Eck, 1967). The entries in this column show the biochemists' preference for abbreviations containing no more than three upper-case letters. These abbreviations have been easy to memorize and convenient to use. Up

Table 1.1. Nucleoside symbols

	1	2	3
Adenosine	A	A	A
N(6)-Dimethyladenosine	DMA	m_2^6A	$\overset{m_2^6}{A}$
1-Methyladenosine	MA	m^1A	$\overset{m^1}{A}$
N(6)-Isopentenyladenosine	NPA	iA	$\overset{i}{A}$
2-Thiomethyl-6-isopentenyladenosine		(sA)	(sA)
Cytidine	C	C	C
5-Methylcytidine	5MC	m^5C	$\overset{m^5}{C}$
N(6)-Acetylcytidine	NAC	acC	$\overset{ac}{C}$
2′-O-Methylcytidine	OMC	Cm	C^m
Guanosine	G	G	G
N(2)-Dimethylguanosine	DMG	m_2^2G	$\overset{m_2^2}{G}$
1-Methylguanosine	1MG	m^1G	$\overset{m^1}{G}$
N(2)-Methylguanosine	2MG	m^2G	$\overset{m^2}{G}$
N(7)-Methylguanosine	7MG	m^7G	$\overset{m^7}{G}$
2′-O-Methylguanosine	OMG	Gm	G^m
Inosine	I	I	I
1-Methylinosine	MI	m^1I	$\overset{m^1}{I}$
Ribosylthymine	T	T	T
Uridine	U	U	U
5,6-Dihydrouridine	DHU	hU	$\overset{h}{U}$
2′-O-Methyluridine	OMU	Um	U^m
Pseudouridine	PSU	Ψ	Ψ
4-Thiouridine		4S	4S
Uridine derivative	U*	$\overset{*}{U}$	$\overset{*}{U}$
Unknown	Y, N, X	N	N
Purine	Pu	R	R
Pyrimidine	Py	Y	Y

to the present, they have been unequivocal in meaning. More recently, a new set of abbreviations has been formulated (see Sober, 1968); they are shown in columns 2 and 3 of Table 1.1. These symbols were suggested by Cohn (1966) and are effective in distinguishing between isomeric nucleosides such as ^{6}N,^{6}N-dimethyladenosine (m_2^6A) and 1-methyl-^{6}N-methyladenosine ($m^{1,6}A$). They are slightly more cumbersome to use, but this should be more than offset by their greater precision in representation. For this reason, Cohn's suggested symbols will be used whenever appropriate in this manual. Column 3 of Table 1.1 illustrates the form that may be used for the individual nucleosides when written as part of a polynucleotide.

In a typical mononucleotide, adenosine-3′-phosphate, the base is linked to ribose (or deoxyribose) by an N′-glycoside bond and the sugar is esterified with phosphate, as shown in Fig. 1.2. The backbone of nucleic acid polymers consists of alternating units of sugar and phosphate; see Fig. 1.3. The bases do not participate in the polymer backbone structure, but instead are attached to the C-1 of the sugars. They interact with each other through hydrogen bonding or secondary stacking forces. An example of a short segment of a polynucleotide is given in Fig. 1.4. This structure may be abbreviated as in Fig. 1.5, where the vertical lines are the sugar moieties, and the diagonal lines are the 3′ → 5′ phosphodiester linkages (with the upper junction of the

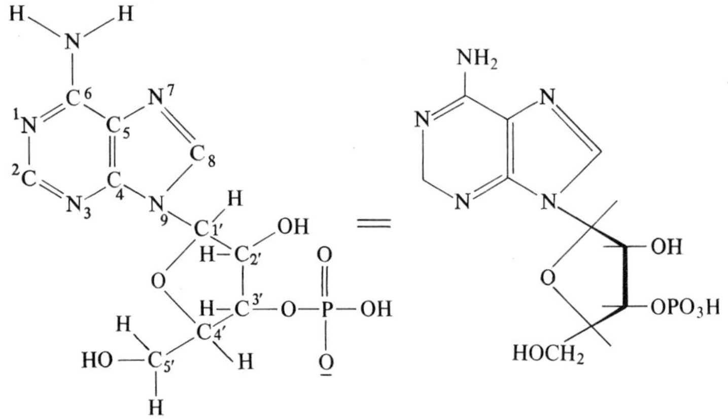

Fig. 1.2. Adenosine 3′-phosphate.

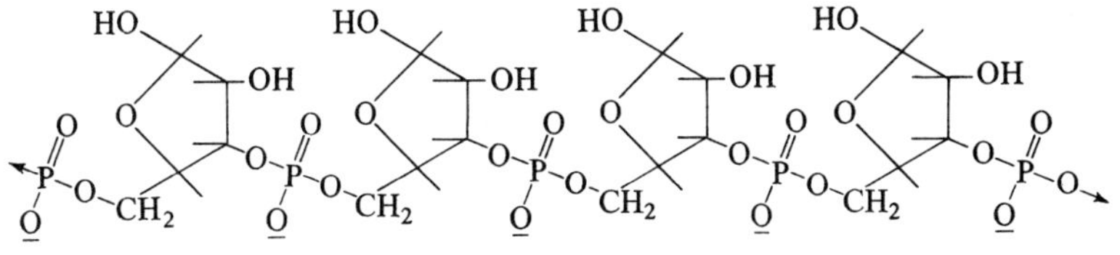

Fig. 1.3. Poly-β-D-ribose phosphate.

diagonal at the 3′ position and the lower junction at the 5′ position). The symbols for the bases are placed on top of the vertical lines.

As the details of the structure of nucleic acids have become more generally known, the need for more compact formulas has increased. The structure in Fig. 1.5 may be shortened even more to the formula . . . pApCpGpUpAp . . . . The upper-case letters represent the nucleosides (sugar and base) and the lower case *p* stands for the phosphoester. The phosphate at the right-hand end of this polynucleotide segment is considered to be esterified to the 3′-hydroxyl of the adenosine which is adjacent at its left. All of the other phosphates represent 3′ → 5′ phosphodiester linkages. According to this convention, pA would represent adenosine 5′-monophosphate (5′-adenylic acid or muscle adenylic acid), and Cp would mean cytidine 3′-monophosphate (3′-cytidylic acid).

The nomenclature becomes a little complicated when segments containing more than one nucleoside are considered. For example, pAp and ppA are both adenosine diphosphates, the former designating adenosine 3′,5′-diphosphate and the latter adenosine 5′-pyrophosphate (adenosine diphosphate or ADP). Fortunately, pyrophosphate linkages are unknown in nucleic acid structures except at the termini. However, even in the absence of pyrophosphate linkages, complications arise when segments containing more than one nucleoside are considered. For example, the structures ApA, pApA, ApAp, and pApAp are all dimers from the standpoint of the number of nucleosides. The first one is a dinucleoside monophosphate and the last one is a dinucleoside triphosphate. The second and third are both dinucleoside diphosphates and cannot be distinguished unless additional information is given concerning the position of the terminal monophosphates. In base sequence work the interest centers mainly around the position of the

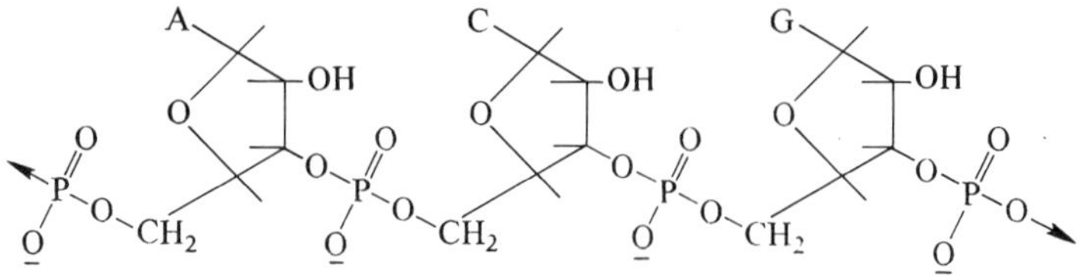

Fig. 1.4. A polynucleotide segment.

bases or nucleosides in the polynucleotide chain. A chemically accurate description of the position of the first base (as we normally read from left to right) in the oligonucleotides ApA, pApA, ApAp, and pApAp is the 3′-hydroxyl-linked end. Similarly, the position of the last base is accurately described as at the 5′-hydroxyl-linked end. These designations would avoid the confusion experienced by the uninitiated as well as some of the initiated in trying to describe the terminal positions of nucleotide chains with free hydroxyl or phosphorylated ends. That there is not more confusion is due to the general acceptance of representing the chain with the 5′-hydroxyl-linked terminus on the right and the 3′-hydroxyl-linked terminus on the left, as shown in the oligonucleotide containing five nucleosides in Fig. 1.5. For simplicity and clarity we shall continue to refer to these termini as the right and left ends.

Abbreviation of nucleic acid structural representations can be carried still further by the substitution of hyphens for the phosphate bond. Accordingly, the segment pApCpGpUpAp would appear as -A-C-G-U-A-. In printing practice, this provides some measure of saving in space. It may also encourage the use of lower-case letters as parts of the symbols for modified bases or nucleosides because there would then be no conflict with the lower-case *p* of the internucleotide bond.

Finally, there is the matter of describing oligonucleotides of various chain lengths, such as diadenosyl triphosphate for pApAp, and tetrauridyl triphosphate for UpUpUpU. These descriptions are adequate and no additional comment is needed. For brevity, however, we often refer to the chain length of an oligonucleotide on the basis of the number of nucleosides present in the molecule. The terms dimer, trimer, tetramer, and so on (derived from the Greek) are used. The use of the Greek-derived terms becomes unwieldy when oligonucleotides

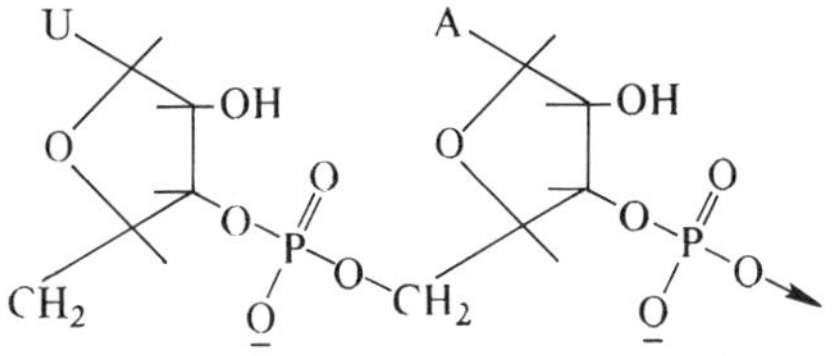

containing 20 to 30 or more nucleosides are being considered. For this reason we shall refer to the shorter oligonucleotides alternatively as pentamer or 5-mer, hexamer or 6-mer, and so on, and to the longer ones exclusively with the numeric prefix, as 70-mer when referring to an oligonucleotide that contains 70 nucleosides. When necessary, or pertinent, the condition of the termini with respect to phosphorylation will be stated explicitly.

CHOICE OF NUCLEIC ACID

In base sequence analysis, there are two important choices to be made at the start: the selection of a nucleic acid and the selection of methods. Very often these choices are related in that one method may be suitable for small molecules exclusively, while another may be the only method workable for nucleic acids of a larger size or particular structure. Apart from this consideration, the suitability of a nucleic acid preparation depends on its purity. It is obviously desirable and virtually necessary to have a molecularly homogeneous nucleic acid, with as few fragments as possible. As little as 5 or 10% of such impurities could be sufficient to make an unequivocal sequence determination impossible. Even in instances where the requirement for

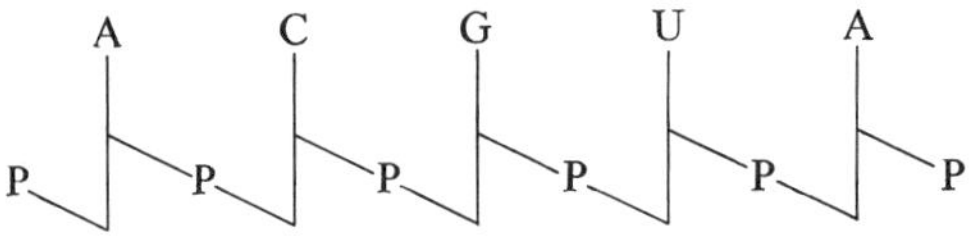

Fig. 1.5. Abbreviated statement of the polynucleotide structure shown in Fig. 1.4.

molecular homogeneity is not so stringent, the purer the nucleic acid, the fewer are the experiments required to resolve equivocal results into a correct sequence.

It is also desirable to obtain a preparation free of nonnucleic acid impurities, such as nucleases or other proteins, that would interfere with subsequent manipulations and interpretation of the data.

Next, it is necessary to obtain homogeneous nucleic acid in adequate amounts. As a rule, the number of equivalents of nucleic acid needed is *at least* equal to the number of equivalents of monomer that can be identified and determined quantitatively. This means that large amounts of high molecular weight nucleic acid are needed, while smaller amounts of low molecular weight nucleic acid are satisfactory. Very often, small amounts of radioactive nucleic acid may be used. Here the requirement is for higher specific radioactivity in the longer nucleic acid.

RIBONUCLEIC ACID (RNA)

At the present the choice of suitable nucleic acids is being made from a very small fraction of the total number of different molecular species available. This group is comprised of the tRNAs, 5S RNA, rRNAs, viral RNA and, in some special circumstances, messenger ribonucleic acid (mRNA). The suitability of nucleic acids in this group is attested to by the fact that complete or partial sequences have been determined for one or more of these types. Table 1.2 lists the nucleic acids for which complete sequences are known (at the time of this writing). These molecules range in chain length from 76 to 85 nucleotides for the transfer RNAs and 120 nucleotides for 5S RNA. Partial sequences are known for many other tRNAs, and presumably these too will be completely determined in a short time.

As one might expect, only partial sequences are known for the larger molecules such as ribosomal and viral RNA. These are listed in Table 1.3. The smallest of these molecules, satellite tobacco necrosis virus (STNV) RNA, contains as few as 1200 nucleotides in its chain. All of the others range in chain length from about 1500 for 16S RNA to 6400 for tobacco mosaic virus (TMV) RNA. The sequences that have been determined represent only a small percentage of the total bases in these molecules. In each of the cases tabulated, the selection of the

Table 1.2. Known RNA sequences

Transfer RNA	Source	Reference
Alanyl	Baker's yeast	Holley *et al.* (1965)
Seryl I, II	Brewer's yeast	Zachau *et al.* (1966)
	Rat liver	Staehlin *et al.* (1968)
Tyrosyl	Baker's yeast	Madison *et al.* (1966)
	su⁺III ϕ80 bacteriophage	Smith *et al.* (1966), Goodman *et al.* (1968)
	T. utilis	Hashimoto *et al.* (1969)
Phenylalanyl	Baker's yeast	RajBhandary and Chang (1968)
	Wheat germ	Dudock *et al.* (1969)
	E. coli	Barrell and Sanger (1969)
Valyl	Baker's yeast	Baev *et al.* (1967)
	T. utilis	Mizutani *et al.* (1968)
	E. coli	Harada *et al.* (1969), Yaniv and Barrell (1969)
Isoleucyl	*T. utilis*	Takemura *et al.* (1969)
Methionyl	*E. coli*	Cory *et al.* (1968), Cory and Marcker (1970)
N-Formylmethionyl	*E. coli*	Dube *et al.* (1968)
5S	*E. coli*	Brownlee *et al.* (1967, 1968)
5S	KB carcinoma cell	Forget and Weissman (1967, 1968)
VA	Adenovirus-infected KB carcinoma cells	Ohe and Weisman (1970)
5S	*E. coli*, chloramphenicol-treated	Jordan *et al.* (1971)
5S	*E. coli* and *Pseudomonas fluorescens*	DuBuy and Weissman (1971)

nucleic acid for sequence analysis grew largely from successful experiences with the purification procedures.

tRNA. Before embarking on the determination of the base sequences in yeast alanyl transfer RNA (alanyl tRNA), Holley and co-workers (Holley *et al.*, 1961, 1963; Apgar *et al.*, 1962) had developed countercurrent distribution purification methods that indicated the feasibility of separating several different transfer RNAs. Countercurrent

Table 1.3. Known partial sequences

	RNA	Region	Reference
16S	*E. coli*	Right end	McIlreavy and Midgley (1967)
	E. coli	Left end	Sugiura and Takanami (1967)
	E. coli	Oligonucleotides	Fellner (1969); Fellner *et al.* (1970)
	Sarcina lutea	Left end	Sugiura and Takanami (1967)
	Bacillus cereus	Left end	Sugiura and Takanami (1967)
	B. subtilis	Left end	Sugiura and Takanami (1967)
	B. stearothermophilus	Left end	Sugiura and Takanami (1967)
	Saccharomyces cerevisiae	Left end	Sugiura and Takanami (1967)
23S	*E. coli*	Right end	McIlreavy and Midgley (1967)
	E. coli	Left end	Takanami (1967)
	E. coli	Oligonucleotides	Fellner (1969); Fellner and Ebel (1970)
	Sarcina lutea	Left end	Sugiura and Takanami (1967)
	B. cereus	Left end	Sugiura and Takanami (1967)
	B. subtilis	Left end	Sugiura and Takanami (1967)
	B. stearothermophilus	Left end	Sugiura and Takanami (1967)
18S	Rabbit reticulocyte	Right end	Hunt (1965)
	L cells	Right end Left end	Lane and Tamaoka (1965)
28S	Rabbit reticulocyte	Right end	Hunt (1965)
	L cells	Left end	Lane and Tamaoka (1965)
Virus	R17	Right end	Dahlberg (1968)
		Left end	Roblin (1968), Adams and Cory (1970)
		Coat protein amino acids	Adams *et al.* (1969)
		Initiation sites, coat protein, protein A, synthetase	Steitz (1969)
		Coat protein termination site	Nichols (1970)

Table 1.3 (continued)

RNA	Region	Reference
MS2	Right end	DeWachter *et al.* (1971)
	Left end	Glitz (1968), DeWachter *et al.* (1968)
f2	Right end	Weith and Gilham (1967), Glitz *et al.* (1968), Dahlberg (1968)
	Ribosome bonding, site (coat protein initiation site)	Gupta *et al.* (1970)
	Coat protein	Nichols and Robertson (1971)
	Left end	Takanami (1966), Dahlberg (1968), Roblin (1968), Ling (1971)
Qβ	Right end	Dahlberg (1968), Weith and Gilham (1969)
	Left end (175 residues)	DeWachter and Fiers (1969), Billeter *et al.* (1969)
	Coat protein initiation site	Hindley and Staples (1969)
TMV	Right end	Steinschneider and Fraenkel-Conrat (1966), Mandeles (1967), Glitz (1968)
	Unique oligonucleotides	Mandeles (1968), Lloyd and Mandeles (1970)
STNV	Right end	Wimmer and Reichmann (1969), Horst *et al.* (1971)
	Left end	Wimmer and Reichmann (1968), Horst *et al.* (1971)

distribution is a technique that employs a two-phase liquid system to fractionate components in a mixture. It depends on the differences in solubility of the components of a mixture in the two immiscible solvents and is essentially an extraction procedure. This has been superseded by column chromatographic and gel electrophoretic separation techniques (see Chapter 10).

The purification of large amounts of alanyl-tRNA was made possible by the development of a method for handling large amounts of yeast

(Holley, 1963). This procedure began with the phenol extraction of 100 lb of yeast in large plastic vats and ordinarily produced 65 g of mixed tRNAs.

5S RNA. In the case of 5S RNA, which occurs as a component of ribosomes, the isolation of large quantities was not feasible. Because of this, Brownlee and Sanger (1967) used radioactive 5S RNA isolated from labeled *E. coli* cells as the material for base sequence analysis. These isolations were performed in either of two ways: sucrose density sedimentation of purified ribosomes in the presence of the detergent sodium dodecyl sulfate (SDS) and lithium chloride, or direct extraction of the RNA from whole cells and fractionation of the extracted RNA by polyacrylamide gel electrophoresis (Hindley, 1967). Little difference was found in the quality of the material obtained by either method; however, because the second one was quicker and easier to carry out, it was the method of choice.

High Molecular Weight RNA. With rRNA and viral RNA, the feasibility of obtaining highly purified preparations again rules the choice of nucleic acid for sequence studies. Both ribosomes and viruses occur as neat little packages of nucleic acid within a protein "bag." These nucleoprotein packages are fairly easily separated from other cellular constituents by various extraction techniques, together with differential and density-gradient sedimentation (see Chapter 10). As a result, these procedures yield a homogeneous population of ribosomes or virus from which an almost equally homogeneous population of nucleic acid molecules may be extracted. It is for this reason that the choice of high molecular weight nucleic acid for sequence determination has been restricted to ribosomal and viral RNA.

From the standpoint of the verification of the genetic code, probably the most interesting RNA that can be used for sequence analysis is messenger RNA. However, virtually no sequence work has been done directly with these molecules, owing to the difficulty of obtaining a molecularly homogeneous preparation. Messengers represent only a small percentage of the total cellular RNA. This small percentage consists of many molecules, perhaps thousands, that vary in size and half-life, and are constantly being turned over at a rapid rate. In addition, attempts to open the cells in order to isolate the mRNA usually disturbs the delicate balance of the cellular constituents and and results in the release of nucleases. In attempts to isolate mRNA, the investigator is confronted at the first step with a mixed population of fully formed, partially formed, and partially degraded RNA that constitutes only a small fraction of the RNA in the system.

At present, with possibly one exception, there are no techniques for the isolation of mRNA comparable in simplicity and success to those for the preparation of ribosomal and viral RNA. The exception may be mRNA for hemoglobin in mammalian (especially rabbit) reticulocytes. These cells have unusual properties: they can be isolated from, and washed free of, plasma; the contents of the cells can be extracted in the presence of a well-controlled medium by alterations in the osmotic pressure; the cells appear to be free of ribonuclease. Though mammalian reticulocytes do not make RNA (Marks *et al.*, 1962; Burny and Chantrenne, 1964), having lost their nuclei in the process of differentiation, they retain for several days the ability to synthesize hemoglobin from the mRNA present in the cell. This strongly implies that the messenger for hemoglobin is long-lived and does not turn over. Consequently, the presence of partially formed or degraded hemoglobin mRNA should be minimal. Because reticulocytes like those from the rabbit make practically no protein other than hemoglobin (Kruh and Borsook, 1956; Dintzis, 1961), it is reasonable to assume that most of the messenger population of the rabbit reticulocyte will consist of hemoglobin mRNA.

Support for this assumption has been obtained from the electron photomicrographs of polyribosome structure from reticulocytes (Slayter *et al.*, 1963). Messenger RNA in these preparations was visualized as the fiber that connects the ribosomes in the polyribosome structure. The length of this fiber was 1500 Å, which is expected for the messenger that codes for a single hemoglobin chain.

As a result of these considerations, Chantrenne *et al.* (1967) have attempted to isolate hemoglobin mRNA through extraction and sedimentation procedures. They were able to obtain an RNA component from reticulocytes that sediments in the region of 9S and is well separated from 16S and 23S rRNA and 4S tRNA. It appears to be completely destroyed when polyribosomes are treated briefly with ribonuclease, and its sedimentation constant of 9S indicates a nucleic acid of around 430 bases (the number required to code for the approximately 145 amino acid residues in a hemoglobin chain). There remains only the task of separating the small number of different hemoglobin mRNAs from each other before base sequence analysis can begin (but see Chapters 8 and 10).

No species of nucleic acid other than those described above have been sufficiently characterized to serve as material for base sequence analysis.

DEOXYRIBONUCLEIC ACID (DNA)

The main difficulty of the use of deoxyribonucleic acid (DNA) directly in base sequence studies stems from the fact that there are no known nucleolytic enzymes for DNA that are as specific in action as those available for RNA. Consequently, methods that depend on the cleavage of nucleic acid at specific points (all of those that have been successful in the sequence analysis of RNA) may not be applied to the determination of base sequences in DNA. However, a choice of DNAs suitable for sequence analysis can be made against the time when methods are developed that are comparable to those currently in use with RNA. A brief survey of the structural features of some DNAs is therefore in order.

Both cellular DNA and viral DNA occur in nature as long, unbranched, polynucleotide chains of deoxynucleoside monomers linked by 3′-5′ phosphodiester bonds. In cells, chromosomal DNA occurs invariably as a duplex, i.e., two strands held together by the hydrogen bonds of complementary bases adenine:thymine (A·T) and guanine: cytosine (G·C). As the complementarity is believed to be virtually complete, any sequence of bases in one chain is exactly paired with the complementary sequence in the other chain. This situation is apparently true not only for chromosomal DNA but also for DNA in other cellular organelles, such as mitochondria, chloroplasts, and centrioles. These cytoplasmic genetic determinants contain duplex DNA which apparently functions like chromosomal DNA but in a fashion largely independent of the nucleus.

Apart from the duplex nature of the strands, the use of cellular DNA for base sequence analysis poses two serious problems. Both of these problems arise from the large size of chromosomal DNA. For example, the DNA from the single "chromosome" of *E. coli* has an estimated molecular weight of 120 × 10^6 daltons. This means that each DNA strand of the duplex contains 200,000 bases. It would be extremely difficult to maintain the integrity of a single, linear polynucleotide with this number of bases. The rate of thermal degradation alone would result in a sufficiently large number of random chain breaks to render the population of chromosomal DNA molecules too heterogeneous for sequence analysis.

In addition, the problem of chemical or biochemical base sequence analysis in a polynucleotide of 200,000 bases is nothing short of monumental. Consider: If a sequence of segments 100 bases in length could be determined each day, it would require over five years of daily determinations (no weekends off, no holidays, no mistakes) in order to establish an entire sequence. It would seem, for the present, that base sequence analysis of chromosomal material is less than promising.

Viral DNA. Viral DNA probably provides a better choice for base sequence analysis than chromosomal DNA from the standpoint of size and ease of preparation. Many viral DNAs are smaller than bacterial chromosomal DNA and easier to prepare undegraded. Some viral DNA is known to occur in circular duplexes of complementary strands, others occur as open-chain duplexes, and still others occur as single-strand circles. There are no known viable viruses composed of open single-strand DNA chains. Table 1.4 lists some of the structural features of a number of viral DNAs with molecular weights below 120 × 10^6 daltons. From the standpoint of size, ϕX174, fd, and the minute virus of mice (MVM) seem to be most suitable. The molecular weight of these viruses is approximately 1.7 × 10^6 daltons and they therefore contain about 5500 bases. This is presumably sufficient to code for six cistrons. Of these, ϕX174 is probably the most convenient to obtain in large quantities. It occurs naturally in the form of a circle and is molecularly homogeneous in all viable viruses. Open, single-strand ϕX174 DNA containing one 3′-hydroxy and one 5′-hydroxy terminus may also be isolated from ϕX174 virus preparations. Although these open DNA chains are of the same molecular weight as the circular ϕX174 DNA, they are not infective and probably represent circles that

Table 1.4. Selected viral DNA structural data

Virus	Host	M.W., daltons $\times 10^{-6}$	Form	Standard Nicks[a]	Strands
ϕX174	*E. coli*	1.7	Circular	—	Single
fd	*E. coli* (male)	1.7	Circular	—	Single
M13	*E. coli* (male)	1.7	—	—	Single
MVM	Mouse	~1.5	—	—	Single
Polyoma	Mammal	~3.0	Circular	no	Duplex
Papilloma	Rabbit	~4.3	Circular	no	Duplex
Papilloma	Bovine	4.9	Circular	—	Duplex
Papilloma	Human	~5.0	Circular	no	Duplex
ϕ15	*B. subtilis*	17	—	—	Duplex
Adeno	Human	~23	Linear	—	Duplex
Tϕ3	*B. stearothermophilis*	23	Linear	—	Duplex
T7	*E. coli*	25	Linear	no	Duplex
P22	*Salmonella*	26	Linear	no	Duplex
λ	*E. coli*	~31	Linear	no	Duplex
T1	*E. coli*	31	Linear	no	Duplex
X	*E. coli* and *Salmonella*	40	—	—	Duplex
SP8	*B. subtilis*	69	—	—	Duplex
T5	*E. coli*	85	Linear	yes	Duplex
SP50	*B. subtilis*	~110	Linear	—	Duplex

[a] "Standard nicks" refers to single-strand breaks at specific, reproducible sites in a population of molecules.

have been broken or "nicked. " Unfortunately, the nicks probably do not occur at any one position in the circle, and hence the chains produced are permutations of each other. This minimizes the usefulness of nicked ϕX174 circles in base sequence determination.

The next larger group of viruses whose structures have been studied includes the mammalian viruses such as polyoma, papilloma, and adeno viruses. At present, they tend to be excluded from sequence analysis because of their characteristically low yields and possible safety considerations.

As we go higher in molecular weight, the prospect of determining base sequences becomes less bright. All of the remaining viruses listed in Table 1.4 are bacterial viruses and range in molecular weight from 17 to 110 $\times$ 10^6 daltons. Of these, P22 illustrates another kind of difficulty that may be encountered. P22 is a bacteriophage for

Salmonella with a molecular weight of 26×10^6 daltons. Although the shape of P22 is a linear duplex, its structure is that of a duplex with a circularly permuted base sequence. An example of circular permutation is the alphabet written in the following five ways:

ABCDEFGHIJKLMNOPQRSTUVWXYZ
BCDEFGHIJKLMNOPQRSTUVWXYZA
CDEFGHIJKLMNOPQRSTUVWXYZAB
DEFGHIJKLMNOPQRSTUVWXYZABC
EFGHIJKLMNOPQRSTUVWXYZABCD

It is obvious that a population of circularly permuted molecules is ill suited for base sequence analysis by most of the available methods. A similar difficulty is posed by viral DNA that is terminally repetitious. An example of a terminally repetitious alphabet is

ABCDEFGHIJKLMNOPQRSTUVWXYZABCDEFG
ABCDEFGHIJKLMNOPQRSTUVWXYZABCD
ABCDEFGHIJKLMNOPQRSTUVWXYZABCDEF
ABCDEFGHIJKLMNOPQRSTUVWXYZABCDEF
ABCDEFGHIJKLMNOPQRSTUVWXYZABC

As illustrated above, terminal repetitions tend not to be of uniform length so that sequence methods that depend on congruence of all of the molecules in a population would not be applicable. The bacteriophage T7 is an example of a virus whose DNA base sequence is terminally repetitious.

In summary, considerations of size, molecular homogeneity, and lack of adequate chemical and biochemical tools tend to make DNA less favorable for base sequence analysis than RNA. In this respect, one possible approach may be through use of viral DNA that may be divided or is naturally divided into subunits. The DNA of T5 bacteriophage is an example. It is a linear duplex of unique sequence with several nicks in specific positions along both chains. Consequently, when T5 DNA is sedimented in alkali (to separate the chains of the duplex), at least six subunits can be recognized in the sedimentation patterns. These subunits range in size from 34 microns to 4 microns. A single-strand subunit 4 microns in length presumably contains about

12,000 nucleotides, still a considerable sequence to determine by present standards.

REFERENCES

Adams, J. M., and Cory, S. (1970). *Nature*, **227**, 570.

Adams, J. M., Jeppesen, P. G. N., Sanger, F., and Barrell, B. G. (1969). *Nature* **223**, 1009.

Apgar, J., Holley, R. W., and Merrill, S. H. (1962). *J. Biol. Chem.*, **237**, 796.

Baev, A. A., Venkstern, T. V., Mirzabekov, A. D., Krutilina, A. I., Li, L., and Axelrod, V. D. (1967). *J. Mol. Biol.*, **1**, 754.

Barrell, B. G., and Sanger, F. (1969). *FEBS Letters*, **3**, 275.

Billeter, M. A., Dahlberg, J. E., Goodman, H. M., Hindley, J., and Weissman, C. C. (1969). *Nature*, **224**, 1083.

Brownlee, G. G., and Sanger, F. (1967). *J. Mol. Biol.*, **23**, 337.

Brownlee, G. G., Sanger, F., and Barrell, B. G. (1967). *Nature*, **215**, 735.

Brownlee, G. G., Sanger, F., and Barrell, B. G. (1968). *J. Mol. Biol.*, **34**, 379.

Burny, A., and Chantrenne, H. (1964). *Biochim. Biophys. Acta*, **80**, 31.

Chantrenne, H., Burny A., and Marbaix, G. (1967). *Prog. Nucl. Acid Res. Mol. Biol.*, **7**, 173.

Cohn, W. E. (1966). *J. Biol. Chem.*, **241**, 527.

Cory, S., and Marcker, K. A. (1970). *Eur. J. Biochem.*, **12**, 177.

Cory, S., Marcker, K. A., Dube, S. K., and Clark, B. F. C. (1968). *Nature*, **220**, 1039.

Dahlberg, J. E. (1968). *Nature*, **220**, 548.

Dayhoff, M. O., and Eck, R. V. (1967). *Atlas of Protein Sequence and Structure.* National Biomedical Research Foundation, Silver Spring, Maryland.

DeWachter, R., and Fiers, W. (1969). *Nature*, **221**, 233.

DeWachter, R., Verhassel, J-P., and Fiers, W. (1968). *FEBS Letters*, **1**, 93.

DeWachter, R., Vandenberghe, A., Merregaert, J., Contreras, R., and Fiers, W. (1971). *Proc. Natl. Acad. Sci.*, **68**, 585.

Dintzis, H. (1961). *Proc. Natl. Acad. Sci.*, **47**, 247.

Dube, S. K., Marcker, K. A., Clark, B. F. C., and Cory, S. (1968). *Nature*, **218**, 232.

DuBuy, B., and Weissman, S. M. (1971). *J. Biol. Chem.*, **246**, 747.

Dudock, B. S., Katz, G., Taylor, E. K., and Holley, R. W. (1969). *Proc. Natl. Acad. Sci.*, **62**, 941.

Fellner, P. (1969). *Eur. J. Biochem.*, **11**, 12.

Fellner, P., and Ebel, J. P. (1970). *Nature*, **225**, 1131.

Fellner, P., Ehresmann, C., and Ebel, J. P. (1970). *Nature*. **225**, 26.

Forget, B. G., and Weissman, S. M. (1967). *Science*, **158**, 1695.

Forget, B. G., and Weissman, S. M. (1968). *J. Biol. Chem.*, **243**, 5709.

Glitz, D. G. (1968). *Biochemistry*, **7**, 927.

Glitz, D. G., Bradley, A., and Fraenkel-Conrat, H. (1968). *Biochim. Biophys. Acta*, **161**, 1.

Goodman, H. M., Abelson, J., Landy, A., Brenner, S., and Smith, J. D. (1968). *Nature*, **217**, 1019.

Gupta, S. L., Chen, J., Schaefer, L., Lengyel, P., and Weissman, S. M. (1970) *Biochem. Biophys. Res. Commun.*, **39**, 883.
Harada, F., Kimura, F., and Nishimura, S. (1969). *Biochim. Biophys. Acta*, **195**, 590.
Hashimoto, S., Miyazaki, M., and Takemura, S. (1969). *J. Biochem.* **65**, 659.
Hindley, J. (1967). *J. Mol. Biol.*, **30**, 175.
Hindley, J., and Staples, D. H. (1969). *Nature*, **224**, 964.
Holley, R. W. (1963). *Biochem. Biophys. Res. Commun.*, **10**, 186.
Holley, R. W., Apgar, J., Merrill, S. H., and Zubkoff, P. L. (1961). *J. Am. Chem. Soc.*, **83**, 4861.
Holley, R. W., Apgar, J., Everett, G. A., Madison, J. T., Merrill, J. H., and Zamir, A. (1963). *Cold Spring Harbor Symp. Quant. Biol.*, **28**, 117.
Holley, R. W., Apgar, J., Everett, G. A., Madison, J. T., Marquisee, M., Merrill S. J., Penswick, J. R., and Zamir, A. (1965). *Science*, **147**, 1462.
Horst, J., Fraenkel-Conrat, H., and Mandeles, S., (1971). *Biochemistry*, **10**, 4748.
IUPAC-IUB CBN (1970). *Eur. J. Biochem*, **15**, 203.
Jordan, B. R., Forget, B. G., and Monier, R. (1971). *J. Mol. Biol.*, **55**, 407.
Jukes T. H. (1970). In *Progress in Nucleic Acid Research*, J. N. Davidson and W. E Cohn (eds.), Vol. 11. Academic Press, New York.
Kruh, J. and Borsook H. (1956). *J. Biol. Chem.* **220**, 905.
Lane, B. G., and Tamaoka, T. (1965). *Biochem. J.*, **95**, 541.
Lewicki, P. P., and Sinskey, A. J. (1970). *Anal Biochem.*, **33**, 273.
Ling, V. (1971). *Biochem. Biophys. Res. Commun.* **42**, 82.
Lloyd D. A. and Mandeles, S. (1970). *Biochemistry*, **9**, 932.
McIlreavy, D. J., and Midgley, J. E. M. (1967). *Biochim. Biophys. Acta*, **142**, 47.
Madison, J. T., Everett, G. A., and Kung, H. K. (1966). *Science*, **153**, 531.
Mandeles, S. (1967). *J. Biol. Chem.*, **242**, 3103.
Mandeles, S. (1968). *J. Biol. Chem.*, **243**, 3671.
Marks, P. A., Burka, E. R., and Schlessinger, D. (1962). *Proc. Natl. Acad. Sci.*, **48**, 2163.
Mizutani, T., Miyazaki, M., and Takemura, S. (1968). *J. Biochem.*, **64**, 839.
Nichols, J. L. (1970). *Nature*, **225**, 147.
Nichols, J. L., and Robertson, H. D. (1971). *Biochim. Biophys. Acta*, **228**, 676.
Ohe, K., and Weissman, S. M. (1970). *Science*, **167**, 879.
RajBhandary, U. L., and Chang, S. H. (1968). *J. Biol. Chem.*, **243**, 398.
Roblin, R. (1968). *J. Mol. Biol.*, **31**, 51.
Slayter, H. S., Warner, J. R., Rich, A., and Hall, C. E. (1963). *J. Mol. Biol.*, **7**, 652.
Smith, J. D., Abelson, J. N., Clark, B. F. C., Goodman, H. M., and Brenner, S. (1966). *Cold Spring Harbor Symp. Quant. Biol.*, **31**, 479.
Sober, H. A. (ed.) (1968). *Handbook of Biochemistry*. The Chemical Rubber Co., Cleveland, Ohio.
Staehelin, M., Rogg, H., Baguley, B. C., Ginsberg, T., and Wehrli, W. (1968). *Nature* **219**, 1363.
Steinschneider, A., and Fraenkel-Conrat, H. (1966). *Biochemistry*, **5**, 2735.
Steitz, J. A. (1969). *Nature*, **224**, 957.
Sugiura, M., and Takanami, M. (1967). *Proc. Natl. Acad. Sci.*, **58**, 1595.
Sugiura, M., Okamoto, T., and Takanami, M. (1969). *J. Mol. Biol.*, **43**, 299.
Takanami, M. (1966). *Cold Spring Harbor Symp. Quant. Biol.*, **31**, 611.
Takanami, M. (1967). *J. Mol. Biol.*, **23**, 135.

Takemura, S., Murakami, M., and Miyazaki, M. (1969). *J. Biochem.*, **65**, 659.
Weith, H. L., and Gilham, P. T. (1967). *J. Am. Chem. Soc.*, **89**, 5473.
Weith, H. L., and Gilham, P. T. (1969). *Science*, **166**, 1004.
Yaniv, M., and Barrell, B. G. (1969). *Nature*, **222**, 278.
Zachau, H. G., Dutting, D., and Feldman, H. (1966). *Z. Physiol. Chem.*, **347**, 212.

2.

CHOICE OF A METHOD

The determination of base sequences in nucleic acids may be carried out by any of a number of biochemical, chemical, or physical procedures. As pointed out before, the choice of a successful method is dependent to a large extent on the overall objective: whether total or partial sequences are required, whether the nucleic acid is of low or high molecular weight. Each of the methods has special features which may be most suitable for a particular objective.

One of the biochemical methods is the overlap method; it involves the reconstruction of nucleic acid sequences from fragments obtained by cleavage of the nucleic acid at specific points. The fragments are fitted together in their original places by virtue of overlapping sequences of bases. Of all methods used in base sequence studies, the overlap method has been by far the most successful. Its application has resulted in the determination of the base sequence in more than 15 different tRNAs and several 5S RNAs (see Table 1.2). Its further application to the study of larger RNA molecules containing more than 150 bases seems to be limited, however, by the increased complexity of the fragments obtained after cleavage of the nucleic acid and by the current state of the analytical art (See Chapters 3 and 4).

A second biochemical method involves the stepwise shortening of nucleic acid with specific exonucleases, such as venom and spleen phosphodiesterases (VPDE and SPDE). The aim of this procedure is to remove and identify one nucleotide at a time from the nucleic acid.

Two main difficulties have been encountered in carrying out this plan: the occurrence of extraneous breaks in the nucleic acid chain due to the presence of endonucleases in the exonuclease preparation or to the basic thermal degradation rate, and the difficulty in establishing and maintaining synchrony for each step (see Chapter 5).

A third method, operationally related to the two above, is the double-label or "handle" method (Mandeles and Tinoco, 1963). It involves the cleavage of nucleic acid that has been labeled at each terminus and the determination of the base sequences in the fragments attached to the label. The presence of the labeled terminus aids in the correct positioning of the various fragments obtained. End modifications have been used successfully to distinguish terminal oligonucleotides from all other oligonucleotides in sequence studies of the larger RNA molecules such as viral RNA (Lee and Gilham, 1966; Weith and Gilham, 1967; Mandeles, 1967; DeWachter and Fiers, 1967; Glitz *et al.*, 1968; Giltz, 1968; Dahlberg, 1968) and ribosomal RNA (Hunt, 1965; Takanami, 1967; Sugiura and Takanami, 1967) (Chapter 7). The limitations of this method appear to be the stability of the labeled nucleic acid during analytical manipulation, and the difficulty of separating long-chain fragments that differ from each other by only a few nucleotides.

The chemical procedure for determining the sequence of bases in nucleic acid is a stepwise process similar to the second biochemical method described above. It involves the amine-catalyzed β-elimination of the right terminal base after periodate oxidation of the vicinal hydroxyls at that end (Whitfeld and Markham, 1953; Brown *et al.*, 1953; Whitfeld, 1954). Additional degradation cycles are initiated by the removal of the phosphate at the newly exposed terminus of the remaining nucleic acid, followed by periodate oxidation and amine-catalyzed β-elimination of the terminal base. This method has been applied successfully to the base sequence determination of the right-hand ends of viral RNA (Steinschneider and Fraenkel-Conrat, 1966; Weith and Gilham, 1967) (Chapter 6). As in the stepwise biochemical method, the chemical method is sensitive to the presence of spurious ends produced as a result of thermal, chemical, or enzymatic breaks.

Of all the methods, one physical method for determining base sequences is potentially the best. It involves the use of electron microscopy to make visible the individual bases that have been rendered dense by combination with suitable derivatives (Beer and Moudrianakis, 1962). In this manner, the base sequences can be determined photographically. At present, improvements in resolution and in the signal-to-noise ratio are needed so that the labeled bases can be distinguished more easily from each other and from the background (Chapter 9).

USEFUL DEGRADATIVE TECHNIQUES

The biochemical and chemical methods described briefly above depend on the execution of a number of common analytical procedures. A preliminary description of these techniques will aid greatly in understanding the various sequence strategies.

All of the biochemical and chemical methods require the specific cleavage of nucleic acids into smaller fragments together with the ability to sort out the products of the cleavage. The most useful procedure for specific degradation of nucleic acids is digestion with pancreatic or T_1 ribonuclease (RNase). Pancreatic RNase specifically cleaves the nucleic acid chains to produce fragments that are terminated by a pyrimidine 3′-phosphate. In a similar manner, T_1 RNase cleaves nucleic acid chains to produce fragments terminated with guanosine 3′-phosphate. This is summarized in the accompanying scheme for the reaction of each nuclease with a hypothetical 10-mer.

ApGpUpCpApUpGpCpApCp

1. $\xrightarrow{\text{Pan RNase}}$ ApGpUp + Cp + ApUp + GpCp + ApCp

2. $\xrightarrow{\text{T}_1\text{ RNase}}$ ApGp + UpCpApUpGp + CpApCp

Studies of the mechanism of these reactions (Sato and Egami, 1957) reveal that they proceed in two steps: first, an attack by the oxygen of the 2′-hydroxyl of the specific nucleoside on the phosphorus of its 3′-phosphate. This attack causes the ribose phosphate chain to break at that point with the formation of a cyclic 2′,3′-phosphate terminus at the right end of each fragment, and a free 5′-hydroxyl at the left. The second step is the hydrolysis of the cyclic 2′,3′-phosphate terminus to a 3′-phosphomonoester. This reaction mechanism is illustrated in Fig. 2.1. In general, the first step is very much faster than the second step,

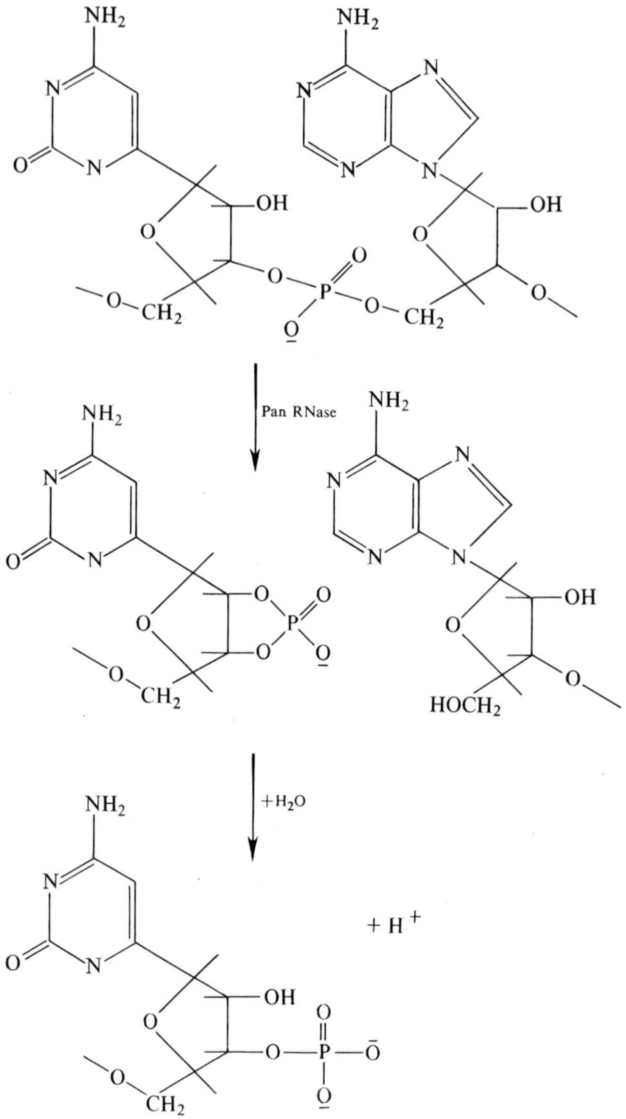

Fig. 2.1. Pancreatic RNase reaction mechanism. Both steps are catalyzed by the enzyme, although the second may also be catalyzed by hydronium ion.

Table 2.1. Calculated oligomer sequence possibilities

Chain Length	Number of Sequences		
n	Random 4^n	T_1 RNase 3^{n-1}	Pan RNase 2^n
1	4	1	2
2	16	3	4
3	64	9	8
4	256	27	16
5	1,024	81	32
6	4,096	243	64
7	16,384	729	128
8	65,536	2,187	256
9	262,144	6,561	512
10	1,048,576	19,683	1,024

so that unless the reaction is allowed to go to completion some of the fragments formed will terminate in cyclic 2′,3′-phosphates.

An idea of the number of different oligonucleotides that may be produced by digestion with pancreatic or T_1 RNase can be obtained by consideration of a very long polynucleotide containing an equal number of essentially random distribution of the four bases. In the case of pancreatic RNase hydrolysis, there are 2^n possible sequences for oligomers with a chain length of n. The reason is that pancreatic RNase distinguishes only between pyrimidines and purines, and the probability of a pyrimidine's occurring at any one position is ½. For T_1 RNase hydrolysis, the probability of a G occurring in any position is ¼ and it can be shown that the number of different oligomers of chain length n produced is 3^{n-1}. If we also consider a completely random hydrolysis of the random polynucleotide, the number of possible oligomers of chain length n produced is 4^n. Table 2.1 lists the values obtained when the indicated calculations are made. It can be seen that the specificity of the two enzymes has reduced the oligonucleotides possible for the shorter chain lengths to an analytically manageable number. The problem of whether an RNase digest of a natural polynucleotide can contain all of the sequence possibilities in any chain length is considered in Chapter 7.

With respect to specificity, a number of different pyrimidine nucleoside phosphate linkages are susceptible to cleavage by pancreatic RNase. They include cytidine, uridine, 5-methylcytidine (but not

2′-O-methylcytidine), dihyrouridine, and pseudouridine. Similarly, in addition to the guanosine 3′-phosphate linkage, T_1 RNase is specific for 2-methylguanosine (but not 7-methylguanosine or 2′-O-methylguanosine), and dimethylguanosine. The rate of enzymatic hydrolysis of dimethylguanosine 2′,3′-cyclic phosphate to the 3′-phosphate is very low relative to the rate for guanosine 2′-3′-cyclic phosphate. As a result, nucleic acids containing dimethylguanosine often yield fragments terminated with a dimethylguanosine 2′,3′-cyclic phosphate even after prolonged digestion with T_1 RNase.

Two enzymes that have been used to identify terminal nucleosides and, in the case of smaller oligonucleotides, to determine sequence, are the exonucleases venom phosphodiesterase (VPDE) and spleen phosphodiesterase (SPDE). The venom enzyme requires that the right-hand end of the oligonucleotide or polynucleotide be unphosphorylated, whereas the spleen enzyme requires that the left-hand end of the molecule be unphosphorylated. When the venom enzyme acts, it produces nucleoside 5′-phosphates, while the spleen enzyme produces nucleoside 3′-phosphates. The release of these monomers from an oligo- or polynucleotide occurs in a stepwise fashion as illustrated with a hypothetical tetranucleoside triphosphate ApGpCpU:

$$\begin{aligned} \text{ApGpCpU} &\xrightarrow{\text{VPDE}} \text{ApGpC} + \text{pU} \xrightarrow{\text{VPDE}} \text{ApG} \\ &\qquad + \text{pC} + \text{pU} \xrightarrow{\text{VPDE}} \text{A} + \text{pG} + \text{pC} + \text{pU} \\ \text{ApGpCpU} &\xrightarrow{\text{SPDE}} \text{Ap} + \text{GpCpU} \xrightarrow{\text{SPDE}} \text{Ap} \\ &\qquad + \text{Gp} + \text{CpU} \xrightarrow{\text{SPDE}} \text{Ap} + \text{Gp} + \text{Cp} + \text{U} \end{aligned}$$

It can be seen that the terminal nucleoside may be identified when the terminals of an oligo- or polynucleotide are unphosphorylated. With phosphorylated ends, however, either the enzyme will be inactive or the terminal nucleoside will be phosphorylated, thus making it indistinguishable from the other nucleotides produced. In this case, treatment of a polynucleotide with a nuclease-free phosphomonoesterase (such as a highly purified alkaline phosphatase from *E. coli* A-19) yields the terminally unesterified polymer which is then susceptible to digestion with either of the exonucleases. Occasionally this extra step may be avoided when the polynucleotide is phosphorylated only at the left terminus. In this instance, digestion with micrococcal nuclease serves to identify each of the termini, as illustrated with the hypothetical oligomer pApGpCpU:

$$\text{pApGpCpU} \xrightarrow{\text{Micrococcal Nuclease}} \text{pAp} + \text{Gp} + \text{Cp} + \text{U}$$

The left terminus is identified by the fact that the nucleoside pAp is doubly phosphorylated, and the right terminus is identified as the simple nucleoside U. The same kind of result may be obtained by the hydrolysis of the oligomer pApGpCpU with alkali. Hydrolysis of RNA acids by alkali is essentially not base-specific; it proceeds by a mechanism similar to that for RNase digestion. With alkaline hydrolysis, however, the intermediate 2′,3′-cyclic phosphate is converted into a mixture of nucleotides with 2′-monophosphate or 3′-monophosphate ends. This procedure has been useful in the determination of base composition of nucleic acids inasmuch as it is reproducible, conditions can be selected so that the side reactions (deamination, dephosphorylation) are minimal, and it is not subject to the occasional uncertainties of enzyme degradation (Bock, 1967) (see Chapter 11).

A number of other procedures have been described for the degradation of nucleic acids (see Grossman and Moldave, 1967). These include the acid hydrolysis and the hydrazinolysis of DNA to produce, respectively, pyrimidine oligonucleotides and purine oligonucleotides. The application of these techniques to base sequence analysis in DNA awaits the development of strategies for determining total sequences.

USEFUL TECHNIQUES FOR THE SEPARATION OF DEGRADATION PRODUCTS

The products of the various degradative procedures will usually differ in chain length, base composition, and sequence, as well as other physical properties. A number of very useful procedures have been devised to separate mixtures of nucleotides and oligonucleotides according to one or more of these properties. Among the more useful methods are chromatography (column, paper, thin-layer), electrophoresis (paper, gel) and sedimentation (linear density gradient, isopycnic). The general theory and practical details of chromatography have been presented in numerous textbooks and review articles.

However, certain aspects of these techniques need to be emphasized here because they serve as integral parts of many sequence strategies.

Column Chromatography. A logical first step in sorting mixtures of oligonucleotides from an enzymatic digest of nucleic acids is the segregation of the molecules strictly according to chain length. The value of any method of this kind depends on its ability to distinguish a molecule of chain length n from $n + 1$. Each chain length separation method displays a characteristic ability to resolve n from $n + 1$ in a limited range: paper electrophoresis is effective up to a chain length of 5; ion-exchange column chromatography up to 20. The separations depend on the fact that an increase of n to $n + 1$ also increases the charge e to $e + 1$. The word isopleth has been used to describe oligonucleotides that contain the same number of nucleosides (from the Greek *isos*, equal + *plethos*, number; Hall and Sinsheimer, 1963).

Column chromatographic separation of oligonucleotide into isopleth fractions is usually carried out on the anion exchangers DEAE-cellulose or DEAE-Sephadex using a salt gradient in the presence of 7 *M* urea. Examples of such separations are shown in Figs. 2.2, 2.3, and 2.4. The elution pattern obtained from the chromatography of an alkaline hydrolysate of polyadenylic acid (poly A) is shown in Fig. 2.2. A partial hydrolysate of this kind yields oligoadenylic acid bearing mixed 2′- and 3′-monophosphate ends. For maximum resolution, these

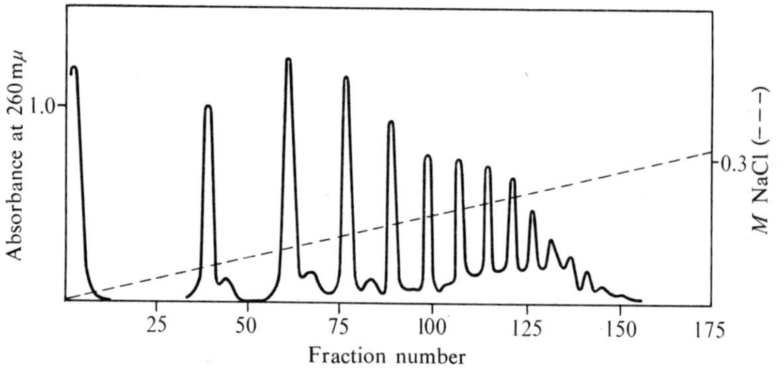

Fig. 2.2. Chromatographic separation of oligo-A on a DEAE-cellulose column. The chromatography is carried out in the presence of 7 *M* urea and a linear salt gradient at pH 7.5.

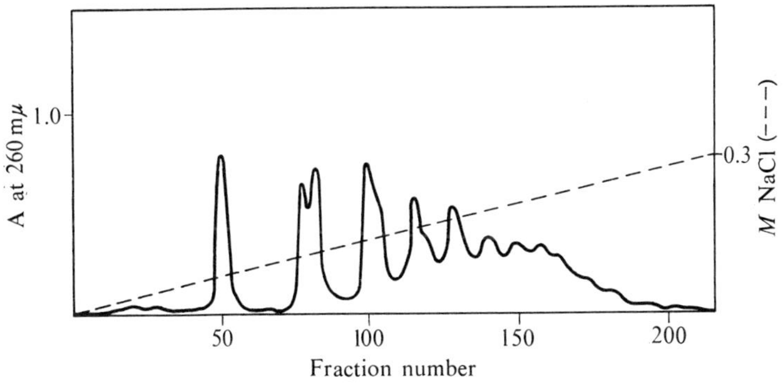

Fig. 2.3. Chromatographic separation of oligonucleotides from a T_1 RNase digestion of TMV RNA. Conditions are the same as in Fig. 2.2.

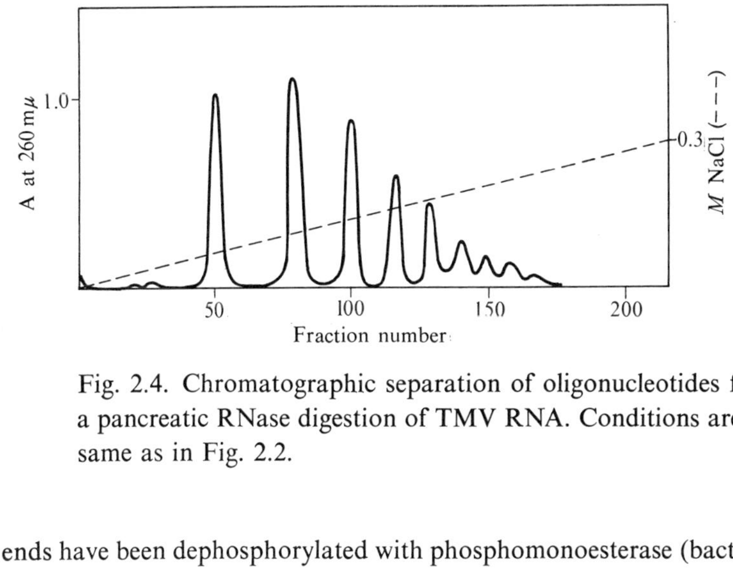

Fig. 2.4. Chromatographic separation of oligonucleotides from a pancreatic RNase digestion of TMV RNA. Conditions are the same as in Fig. 2.2.

ends have been dephosphorylated with phosphomonoesterase (bacterial alkaline phosphatase), and the chromatography is carried out at pH 7.5, where the negative charge of the phosphate group is predominant (Ishikura *et al.*, 1966). The first peak to emerge from the DEAE-cellulose is uncharged adenosine (A), formed from the dephosphorylation of the mixed 2′- and 3′-adenylic acid. The second peak consists of the singly charged diadenosyl monophosphate, ApA, and subsequent peaks contain other oligoadenylic acids of the general formula (Ap)nA.

Under optimal chromatographic conditions, it is possible to achieve the resolution of oligonucleotides into isopleths containing up to 12 or more nucleoside residues (Khorana and Vizsolyi, 1961).

In the case of natural nucleic acid, the resolution tends to become blurred as the base compositional complexity increases. Figures 2.3 and 2.4 show the elution pattern obtained from the chromatography of enzymatic hydrolysis products of tobacco mosaic virus RNA (TMV RNA). It can be seen that oligonucleotides from a T_1 RNase digest (Fig. 2.3) containing the mixed sequences of A, U, and C with a terminal G are less well resolved than oligonucleotides obtained from a pancreatic RNAse digest that essentially are sequences of purines terminated by a pyrimidine (Fig. 2.4).

Apparently there are sufficient forces between oligonucleotides and the DEAE-cellulose other than the phosphate ion-exchange interaction that effect a resolution according to base composition at the expense of chain length resolution. This may be due to the presence of a small percentage of carboxyl groups in the cellulose which, in effect, serve as cation exchangers, and which may also account for the increased ability of DEAE-cellulose to resolve according to base composition after being washed in alkali, a procedure that results in an increase in carboxyl content from air oxidation.

In order to circumvent this difficulty, Rushizky *et al.* (1964) introduced the use of DEAE-Sephadex as a column material for separation of oligonucleotides. DEAE-Sephadex contains fewer carboxyl groups, and therefore its use is more likely to result in the separation of oligonucleotides simply by chain length. That an improvement is in fact obtained can be seen by comparing the pattern of Fig. 2.3 with that of Fig. 2.5, which is a pattern obtained from the chromatographic separation of a T_1 RNase digest of TMV RNA on DEAE-Sephadex.

Elution with Linear Gradients. The effect of changes in salt concentration is to elute the oligonucleotides according to their total charge: those with the smallest charge are the least tightly bound and are eluted at the lowest salt concentration; those with the greater charge are more tightly bound and are eluted at higher salt concentrations. A number of investigators have observed a linear relationship between the log of the charge of the oligonucleotide and the concentration of

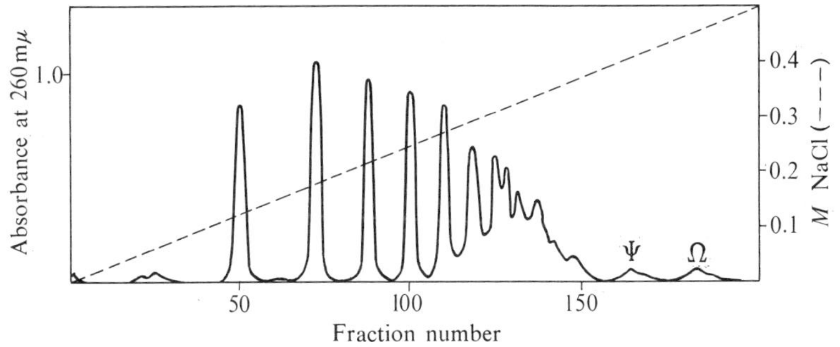

Fig. 2.5. Chromatographic separation of oligonucleotides from a T_1 RNase digestion of TMV RNA on a DEAE-Sephadex column. The chromatography conditions are essentially the same as specified in Fig. 2.2 except that the salt gradient is steeper to accommodate the higher capacity of DEAE-Sephadex.

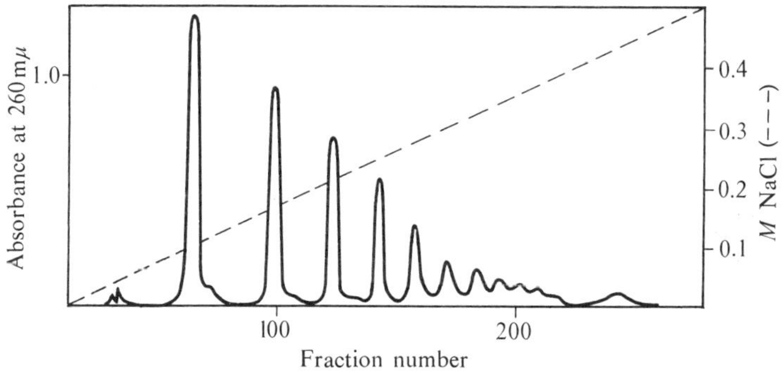

Fig. 2.6. Chromatographic separation of oligonucleotides from a T_1 RNase digestion of rRNA. Conditions are the same as in Fig. 2.5.

salt in the eluting solvent. This is illustrated in Figs. 2.5 and 2.6. Figure 2.6 is the pattern obtained from the separation of rRNA oligonucleotides on DEAE-Sephadex. The oligonucleotides were produced by hydrolysis of the rRNA with T_1 RNase. A plot of the log of charge versus the concentration of the eluting solvent is shown in Fig. 2.7. It can be seen that, for oligonucleotides with charges up to 10, the relationship is linear. Similar observations have been made when

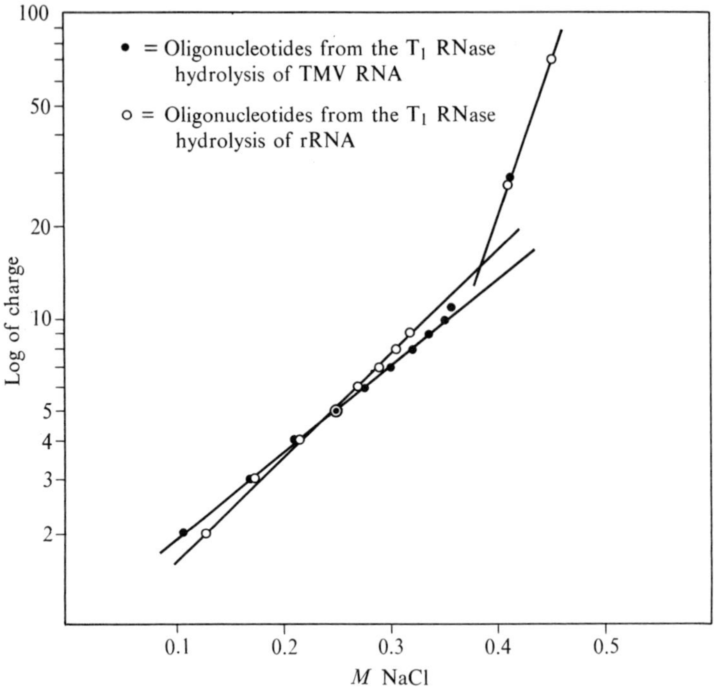

Fig. 2.7. Log of charge versus the concentration of eluting solvent for chromatographic separations on DEAE-Sephadex in the presence of 7 *M* urea.

DEAE-cellulose has been used to achieve chain length separations, especially in the case of oligonucleotides from homopolymers. A theoretical basis for this phenomenon has been discussed by Vermeulen and Hiester (1952) and Lloyd (1969).

An apparent deviation from this rule is observed with T_1 RNase oligonucleotides of chain length greater than 15, as shown in Fig. 2.5. The last two peaks in this pattern have been designated Ψ and Ω. The Ψ peak contains two components, Ψ_1 and Ψ_2, and the Ω peak contains only one component (Mandeles, 1968). The chain length of each Ψ-mer is 26, and that of the Ω-mer is 70 (Lloyd and Mandeles, 1970). When the log ε of these oligonucleotides is plotted against salt concentration in Fig. 2.7, the values fall on a curve whose slope is significantly greater than that for the shorter oligonucleotides. The intersection of the two curves corresponds to a chain length of 15. It is as if the longer oligonucleotides had been chromatographed on a column of smaller ion-exchange capacity.

This effect is not altogether surprising in view of the structure of DEAE-Sephadex. The pore size of the Sephadex beads presumably permits the free diffusion of molecules of molecular weight less than 5000 while restricting those of higher molecular weight to interaction at the surface only. This could account for the two-column effect since the smaller oligonucleotides would interact with the total ion-exchange capacity of the DEAE-Sephadex while the larger oligonucleotides would be restricted to the ion-exchange capacity at the surface of the beads.

Temperature Effects. The effect of changes in temperature on chromatographic separations usually may be attributed to temperature-dependent changes in the rate of diffusion of solute to and from the ion-exchange site, and on the kinetics of the ion-exchange reaction. With the chromatography of nucleic acids, however, a quantitatively greater effect may be due to the temperature-dependent changes in conformation of the polynucleotide chain: as the temperature is increased, the polynucleotide is melted out and more sites are made available for ion exchange. This should result in stronger binding between the polynucleotide and the ion exchanger and, as a result, higher salt concentrations would be required to elute the polynucleotide than would be necessary at lower temperatures. This effect was observed by Baguley *et al.* (1965) in the separation of a mixture of tRNAs on DEAE-cellulose at 20°, 40°, 60°, and 65°. The chromatographic patterns obtained at these different temperatures are shown in Fig. 2.8. In addition to the requirement for higher salt concentrations at the higher temperatures, there appears to be a resolution of the mixed tRNAs into distinct components at 65°. This temperature is close to a mean T_m for most tRNAs. A similar resolution at 20° is not apparent. Assays for amino acid-accepting activity of the peaks confirmed the increased resolution at 65°.

If the major effect of an increase in temperature is to melt out and increase the binding of the longer-chain oligonucleotides, the effect on

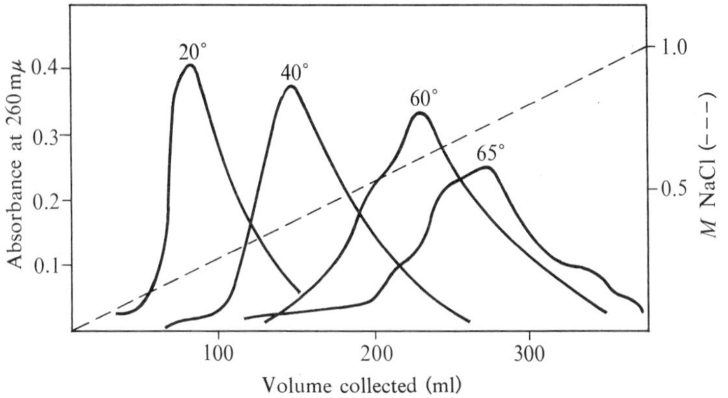

Fig. 2.8. Effect of temperature on elution pattern of tRNA (redrawn from Baguley *et al.*, 1965). Separation of mixed tRNAs performed on DEAE-cellulose columns at temperatures shown.

the shorter oligonucleotides should be less. This would tend to counteract the apparent limitation in distinguishing n from $n + 1$ for the longer oligonucleotides.

Base Composition Effects. Further separation of the individual members of an isoplethic fraction into sequence or base composition isomers may be accomplished through use of anion-exchange column chromatography at low pH. In this case, conditions are deliberately chosen so that the phosphate is essentially uncharged and the binding forces arise instead from the interaction of the bases with the exchanger. Dowex-1 (AG 1) is particularly effective for this purpose because it is an amine-substituted polystyrene: the styrene moiety interacts with the purine and pyrimidine bases through van der Waals forces. Because the charge of all of the bases can be altered drastically between pH 2 and 4, separations by base composition or sequence are usually accomplished at low pH.

Fingerprints of Oligonucleotides. The ability to separate oligonucleotides by chain length and base composition in the same analysis is provided by the fingerprint technique. In one example of this method (Rushizky and Knight, 1960; Rushizky and Sober, 1962) a mixture of oligonucleotides is first separated on paper electrophoreticaly according to base composition. This separation is accomplished at low pH, and the components of the oligonucleotide mixture are spread out along one edge of a sheet of filter paper. The paper is rotated 90° and the chromatogram is developed in a mixed aqueous and organic solvent so that the components of the band now migrate according to chain length at right angles to the electrophoresis direction. In this fashion, separations of oligonucleotides have been obtained up to a chain length of 4 or 5.

The result is an array of spots as shown in Fig. 2.9. This pattern is fairly reproducible and, once the composition of a particular spot has been determined, subsequent identifications can be made quite easily. For the identification of sequence isomers, separations of this kind are ordinarily inadequate because few or no separations of sequence isomers are obtained on paper.

Similar fingerprints of oligonucleotide mixtures may be obtained with thin-layer cellulose. Two kinds of separations have been carried out: thin-layer electrophoresis followed by thin-layer chromatography in a manner analgous to the procedure described above for paper (Bergquist, 1965) and thin-layer chromatography in both directions (Weimann and Randerath, 1963; Randerath and Weimann, 1963). In general, the separations of oligonucleotides are better when one of the two dimensions is an electrophoresis, perhaps because of the better control of the state of ionization of the bases.

Probably the most powerful procedure available for the separation of the shorter oligonucleotides is the fingerprint method developed by Sanger *et al.* (1965). In this procedure, oligonucleotides are first separated by electrophoresis on a strip of cellulose acetate. Next, the material on the cellulose acetate strip is transferred to a sheet of DEAE-paper (a critical manipulation) and the oligomers are further separated by electrophoresis at low pH. When this procedure is carried out correctly, separations are obtained of *sequence* isomers with chain lengths up to 4. After preliminary calibration experiments with oligonucleotides of known sequence, the sequence of an unknown oligomer may be determined from its position in the two-dimensional pattern only. In addition, through use of radioactive nucleic acid only very small amounts of material are needed and consequently large-scale preparations can be avoided. This method was employed initially to determine the complete primary structure of 5S RNA in *E. coli* by

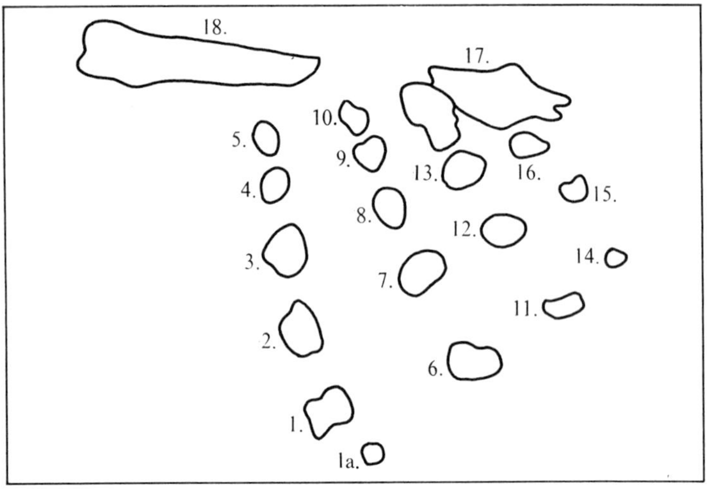

Fig. 2.9. Fingerprint of an oligonucleotide mixture on filter paper. Electrophoresis carried out at low pH for the first dimension followed by chromatography in a mixed aqueous-organic solvent in the second dimension. This is a tracing of the fingerprint reported by Rushizky and Sober (1962). Some of the oligonucleotides were identified as follows:

1a. G!
1. Gp
2. ApGp + CpGp
3. $(Ap)_2Gp + (Cp)_2Gp +$
4. $(Ap)_3Gp + (Cp)_3Gp +$ (Ap, 2Cp)Gp + (2Ap, Cp)Gp
5. Not identified
6. UpGp
7. (CpUp)Gp
8. (2Ap, Up)Gp

Brownlee *et al.* (1967, 1968) and of 5S RNA from KB cells by Forget and Weissman (1967), and since has been used for tRNA and virus RNA. (See chapter 4 for the details of this method.) The following description of the base sequence analysis of two oligonucleotides illustrates the way these analytical procedures are used. The examples chosen are from the studies of RajBhandary and Chang (1968) on the structure of phenylalanine tRNA. They will also be discussed later in connection with a description of the overlap sequence method.

Identification of an Oligonucleotide from a T_1 RNase Digestion. Phenylalanine tRNA from yeast was hydrolyzed completely with T_1 RNase. After removal of the enzyme, the products of the digestion were separated by chromatography on a column of DEAE-cellulose in 7 *M* urea (Tomlinson and Tener, 1963). A linear gradient of sodium chloride was used to elute 18 different fractions, each of which was isolated and analyzed in turn. One of the later fractions, on further chromatography on paper in an *n*-propanol-ammonia-water solvent, was shown to contain two components. The faster-moving component, as a result of combined degradation with pancreatic RNase and phosphomonoesterase (PME), gave 2 moles of cytidine and 1 mole of uridine. It also yielded 1 mole of a dinucleoside monophosphate which was identified as ApG from paper chromatography in a 2-propanol-ammonia-water solvent. Because this oligonucleotide was originally derived from a T_1 RNase digestion, and therefore must end in G, the ApG had to be the right-hand terminal sequence of the fragment. The left terminal was determined by treating some of the original fragment with PME, followed by digestion with snake venom phosphodiesterase (VPDE). As a result of this treatment, a mole each of cytidine, pC, pU, pA, and pG, was produced (VPDE attacks from the right and produces 5′-nucleotides). Cytidine, then, was identified as the left terminal nucleoside, thus permitting the deduction of Cp(Up,Cp)ApGp as the partial sequence of the fragment. The relative order of the nucleotides within the parentheses was established after a partial digestion of the fragment with VPDE. Among the products of this treatment was a trinucleoside diphosphate which (by virtue of the mechanisms of VPDE action) must include the first three nucleosides at the left terminal of the original pentanucleoside tetraphosphate. The trinucleoside diphosphate was isolated, treated with pancreatic RNase, and the products separated by paper chromatography in a 2-propanol-ammonia-water solvent. The products of the last step were Cp, Up, and cytidine. These products could only be obtained with pancreatic RNase if cytidine was on the right-hand end of the trinucleoside diphosphate. It was known, however, from the evidence with VPDE, that a cytidine is the nucleoside at the left terminus, so that the trinucleoside diphosphate must be CpUpC and the original pentanucleoside pentaphosphate must therefore be CpUpCpApGp.

Identification of an Oligonucleotide from a Pancreatic RNase Digestion. When phenylalanine tRNA was hydrolyzed with pancreatic RNase and the products of the digestion were separated on a DEAE-cellulose column with ammonium carbonate as eluting solvent, 16 major fractions were obtained. One of these, fraction 12, was closely associated with the neighboring fraction 13 in a double peak. The fractionation "cut" was made at the minimum point between the two peaks, and the terminal phosphates were removed from the oligonucleotides in fraction 12 by treatment with PME. Paper chromatography of this fraction in an isobutyric acid-ammonia-water solvent revealed the presence of three components: two major components only partially separated from each other, but well separated from the third minor component (which was later shown to be part of fraction 13). A complete separation of the two major components was accomplished only after prolonged paper chromatography (68 hours) in an ethanol-1 *M* ammonium acetate solvent. One of the two components fluoresced strongly under ultraviolet irradiation and proved to be a hexanucleotide containing the minor nucleosides 2′-O-methylguanosine and pseudouridine at the left and right ends, respectively, in addition to the unidentified, fluorescent base Y in the interior. This oligonucleotide proved to be difficult (but not impossible) to analyze because of the absence of any phosphodiester linkages hydrolyzable by T_1 RNase and its relative resistance to exonuclease action.

The second component, a pentanucleotide, was characterized by degradation with T_1 RNase and paper chromatography of the products in *n*-propanol-ammonia-water solvent to yield a dimer, ApGp, and a trimer. The trimer was digested with VPDE to yield adenosine, pC^m (2′-O-methylcytidine-5′-phosphate), and pU. This established adenosine as the nucleoside at the left end of the trimer. Alkaline hydrolysis of the trimer yielded Ap and the dimer C^mpU (separated in a 2-propanol-ammonia-water solvent). The structure of the trimer was therefore ApC^mpU. The ApGp obtained from T_1 RNase action on the original pentanucleotide could only be fitted on the left hand end to give the structure

$$ApGpApC^mpUp$$

In many instances, monomers, dimers, and trimers may be identified, without degradation, merely by their positions on chromatograms, their ultraviolet absorption spectra, or their optical rotatory dispersion spectra. For the larger oligonucleotides, however, procedures more or less as complicated as those described are necessary.

REFERENCES

Baguley, B. C., Bergquist, P. L., and Ralph, R. K. (1965). *Biochim. Biophys. Acta.*, **95**, 510.

Beer, M., and Moudrianakis, E. N. (1962). *Proc. Natl. Acad. Sci.*, **48**, 409.

Bergquist, P. L. (1965). *J. Chromatog.*, **19**, 615.

Bock, R. M. (1967). In *Methods in Enzymology*. L. Grossman and K. Moldave (eds.), Vol. 12, part A, p. 224. Academic Press, New York.

Brown, D. M., Fried, M., and Todd, A. R. (1953). *Chem. Ind.*, 352.

Brownlee, G. G., Sanger, F., and Barrell, B. G. (1967). *Nature*, **215**, 735.

Dahlberg, J. E. (1968). *Nature*, **220**, 548.

DeWachter, R., and Fiers, W. (1967). *J. Mol. Biol.*, **340**, 507.

Forget, B. G., and Weissman, S. M. (1967). *Science*, **158**, 1695.

Glitz, D. G. (1968). *Biochemistry*, **7**, 927.

Glitz, D. G., Bradley, A., and Fraenkel-Conrat, H. (1968). *Biochim. Biophys. Acta*, **161**, 1.

Grossman, L., and Moldave, K. (eds.) (1967). *Methods in Enzymology*, Vol. 12, part A. Academic Press, New York.

Hall, J. B., and Sinsheimer, R. L. (1963). *J. Mol. Biol.*, **6**, 115.

Hunt, J. A. (1965). *Biochem. J.*, **95**, 541.

Ishikura, H., Naelon, F. A., and Cantoni, G. L. (1966). *Science*, **153**, 300.

Khorana, H. G., and Vizsolyi, J. P. (1961). *J. Am. Chem. Soc.*, **83**, 675.

Lee, J. C., and Gilham, P. T. (1966). *J. Am. Chem. Soc.*, **88**, 5685.

Lloyd, D. (1969). Ph.D. Thesis, Dept. of Chemistry, University of California.

Lloyd, D., and Mandeles, S. (1970). *Biochemistry*, **9**, 932.

Mandeles, S. (1968). *J. Biol. Chem.*, **243**, 3671.

Mandeles, S., and Tinoco, I., Jr. (1963). *Biopolymers*, **1**, 183.

RajBhandary, U. L., and Chang, S. H. (1968). *J. Biol. Chem.*, **243**, 598.

Randerath, K., and Weimann, G. (1963). *Biochim. Biophys. Acta*, **76**, 129.

Rushizky, G. W., and Knight, C. A. (1960). *Virology*, **11**, 236.

Rushizky, G. W., and Sober, H. A. (1962). *J. Biol. Chem.*, **237**, 834.

Rushizky, G. W., Bartos, E. M., and Sober, H. A. (1964). *Biochemistry*, **3**, 626.

Sanger, F., Brownlee, G. G., and Barrell, B. G. (1965). *J. Mol. Biol.*, **13**, 373.

Sato, K., and Egami, F. (1957). *J. Biochem.*, **44**, 753.

Steinschneider, A., and Fraenkel-Conrat, H. (1966). *Biochemistry*, **5**, 2735.

Sugiura, M., and Takanami, M. (1967). *Proc. Natl. Acad. Sci.*, **58**, 1595.

Takanami, M. (1967). *J. Mol. Biol.*, **23**, 135.

Tomlinson, R. V., and Tener, G. M. (1963). *Biochemistry*, **2**, 697.

Vermeulen, T., and Hiester, N. K. (1952). *Ind. Eng. Chem.*, **44**, 636.

Weimann, G., and Randerath, K. (1963). *Experientia*, **19**, 49.

Weith, H. L., and Gilham, P. T. (1967). *J. Am. Chem. Soc.*, **89**, 5473.

Whitfeld, P. R., and Markham, R. (1963). *Nature*, **171**, 1151.

Whitfeld, P. R. (1954). *Biochem. J.*, **58**, 390.

3.

THE OVERLAP METHOD AND tRNA

The task of determining amino acid sequence in proteins seemed to be, at its inception, as formidable as base sequence analysis of nucleic acids now appears. Nevertheless, Sanger and Tuppy (1951) worked out the entire primary structure of insulin largely by application of the overlap method. This method employs the specific degradation of the polymer in at least two different ways to produce two nonidentical lists of subunits. In the case of proteins, specific degradations are carried out with specific proteolytic enzymes. The structures of the subunits from one list are compared with the structures of the other list. Sequences that are uniquely common to subunits from each list overlap and permit reconstruction of larger subunits from smaller ones. This can be illustrated with the hypothetical polymer ABAACDDCBA, where A, B, C, and D are monomers. Two enzymes, one that splits bonds to the right of B only and the other that splits bonds to the right of C only, produce two lists of subunits. The B lists, produced by the enzyme, contains the subunits or fragments A, AB, AACDDCB. The C list, produced by the C enzyme, contains the fragments BA, DDC, ABAAC. This is illustrated in Table 3.1.

It is always useful to identify the terminal subunit before searching for overlapping sequences. This can be done as follows: in general, the internal subunits produced by the B enzyme should end in B, and the internal subunits produced by the C enzyme should end in C. In both

Table 3.1. Overlapping lists of fragments

B Enzyme	C Enzyme
AB AACDDCB A	ABAAC DDCBA
A	BA
AB	DDC
AACDDCB	ABAAC

the B and C lists, there is one subunit that ends in A. This can only occur if A is the right-hand terminal monomer. Further, in list C, this terminal subunit is identified as BA. Since BA occurs in the C list, the terminal sequence is CBA (in order to have a terminal BA in the C list, C must be to the left of BA). A similar determination of the left terminal can be made only with agents that split on the left side of specific monomers.

In list C, the overlapping sequence from the B list determines which of the two fragments, DDC and ABAAC, precedes the BA. It can be seen that in the B list the largest subunit AACDDCB contains the sequence AAC. This corresponds uniquely to a portion of the ABAAC subunit from the C list. If these regions of overlap, ABAAC and AACDDCB, are combined, the fragment ABAACDDCB is obtained. This leaves only one fragment on the B list, the single A, to be accounted for. Because the remaining A must be the right-hand terminal, and in fact is part of a CBA terminal sequence, the original sequence must be reconstructed as ABAACDDCBA.

The overlap method of base sequence analysis has been successful in establishing the primary structure of at least 15 tRNAs as well as 5S RNA from two different sources. Application of this method involves the complete digestion of RNA by the specific nucleases, T_1 and pancreatic RNase, identification (by base sequence determination) and quantitation of each oligomer produced, and comparison of T_1 RNase oligomers with pancreatic RNase oligomers for common sequences. Except in the case of small oligomers, this procedure rarely results in a unique reconstruction of the RNA. Usually, additional areas of overlapping sequences must be obtained from partial hydrolysis of the RNA by T_1 and pancreatic RNase. An example of the assembly and comparison of such lists is given for yeast alanyl tRNA, the base nucleoside composition of which is shown in Table 3.2. The products

Table 3.2. Nucleoside composition of yeast alanyl transfer RNA*

23-C	(Cytidine)	1-U*	(Mixture of U and hU)
7-A	(Adenosine)	1-T	(Ribothymidine)
25-G	(Guanosine)	1-m^1G	(1-Methylguanosine)
11-U	(Uridine)	1-m^2_2G	(N^2-dimethylguanosine)
2-hU	(5,6-Dihydrouridine)	1-I	(Inosine)
2-Ψ	(Pseudouridine)	1-m^1I	(1-Methylinosine)

* Adapted from Holley *et al.*, 1965

Table 3.3. Oligonucleotides obtained from a T_1 RNase digest of yeast alanyl tRNA

pG-	2 A-G-	$\overset{h}{U}$-A-G-	U-C-C-A-C-C-OH
9 G-	U-G-	C-$\overset{m^1}{I}$-Ψ-G-	U*-C-U-C-C-G-
C-$\overset{m^2_2}{G}$	U-A-G-	T-Ψ-C-G-	A-U-U-C-C-G-
U-$\overset{m^1}{G}$-	$\overset{h}{U}$-C-G-	A-C-U-C-G-	C-U-C-C-C-U-U-I-
4 C-G-			

NOTE: The hyphens, -, indicate a phosphate.

Table 3.4. Oligonucleotides obtained from a pancreatic RNase digest of yeast alanyl tRNA

C-OH	$\overset{m^1}{I}$-Ψ-	A-G-C-	G-G-$\overset{h}{U}$-
13 C-	$\overset{m^2_2}{G}$-G	A-G-$\overset{h}{U}$-	G-G-A-C-
Ψ-	2 G-C-	G-A-U-	pG-G-G-C-
6 U-	4 G-U-	I-G-C-	G-G-G-A-G-A-G-U*-
A-C-	$\overset{m^1}{G}$-G-C-	G-G-T-	

obtained from T_1 and pancreatic RNase digestions of yeast alanyl tRNA are shown in Tables 3.3 and 3.4 (Holley *et al.*, 1965).

It can be seen in Table 3.2 that there are 8 nucleosides other than adenosine, cytidine, guanosine, and uridine in yeast alanyl tRNA. At first the modified nucleosides complicated the sequence analysis because their presence was not expected and their identification proved to be time-consuming. In addition, some of them were chemically altered in the alkaline or acidic conditions used in the analysis. For example, in alkali, dihydrouridine and 7-methylguanosine are degraded, and 1-methyladenosine is isomerized to 6-methyladenosine. It is for this reason that tRNA is now ordinarily hydrolyzed to completion with VPDE (Hall, 1965) in order to obtain a base composition. An excellent summary of the properties of the modified nucleoside constituents of nucleic acids is given in the *Handbook of Biochemistry* (Sober, 1968). A more complete treatment can be found in Hall (1971). In Table 3.3, most of the fragments produced by T_1 RNase end in guanosine-3′-phosphate (G-). With one exception, the remaining fragments end in purine nucleotides that are either derivatives of G (m^1G- or m_2^2G-) or are closely related structurally (Chapter 1). The one exception is the fragment U-C-C-A-C-C-OH. This fragment does not end in a nucleotide for which T_1 is specific; therefore it represents the terminal oligonucleotide at the right end of alanyl tRNA. Furthermore, the presence of a cytosine nucleoside at the terminus rather than an adenosine nucleoside accords with the fact that most tRNAs isolated from commercial baker's yeast lack the terminal adenosine. The presence of a cytidine residue in the list of pancreatic RNase digestion products confirms the position of cytidine at the right-hand terminus as shown in Table 3.4.

The identity of the left-hand terminus is indicated by the presence of pGp in the T_1 RNase list. Considering the strict specificity of T_1 RNase for producing 3′ phosphorylated ends, the nucleotide bearing a 5′-phosphate could only come from the left-hand end. It may be seen in Table 3.4 that a single fragment bears this 5′-phosphate and that this in fact identifies the fragment pG-G-G-C- as the left terminus.

Next, each list may be searched for oligonucleotides with overlapping sequences. For example, U*- is one of the "odd" nucleotides in alanyl tRNA and occurs only in the whole molecule (as indicated, U* is actually a mixture of U and hU). The only oligonucleotide that contains U*- in the pancreatic digest (Table 3.4) is G-G-G-A-G-A-G-U*-. Similarly, U*- is present in the T_1 RNase fragment U*-C-U-C-C-G- (Table 3.3). These two oligonucleotides overlap at the position of U*-, and the combined structure is therefore G-G-G-A-G-A-G-U*-C-U-C-C-G-.

Table 3.5. Sequences that account for all the nucleotide residues in alanyl tRNA

pG-G-G-C	C-U-C-C-C-U-U-I-G-C-
G-U-G-	$\overset{m^1}{\text{I}}$-Ψ-G-
U-$\overset{m^1}{\text{G}}$-G-C-	G-G-G-A-G-A-G-U*-C-U-C-C-G-
G-C-G-U-A-G-	G-T-Ψ-C-G-
$\overset{h}{\text{U}}$-C-G-	A-U-U-C-C-G-
G-$\overset{h}{\text{U}}$-A-G-	G-A-C-U-
C-G-	C-G-
C-$\overset{m_2^2}{\text{G}}$-G-	U-C-C-A-C-C-A-OH-

NOTE: The terminal fragment includes the final nucleoside adenosine.

By matching oligonucleotides with overlapping sequences from each list, a combined list may be assembled that accounts for all of the nucleotide residues in alanyl tRNA (Holley *et al.*, 1965). This is shown in Table 3.5. In this list, there are 14 oligonucleotides exclusive of the two terminal fragments that can be fitted together in 14 (!) ways, or approximately 87×10^9 permutations. Obviously, additional overlapping areas must be obtained if the correct sequence is to be reconstructed. This was in fact accomplished by analysis of the products of a limited digestion of the tRNA. In the case of alanyl tRNA, digestion with T_1 RNase was carried out for a very brief period at 0° (Penswick and Holley, 1965). As a result, two large fragments were isolated representing the two complete halves of the original molecule. This is shown in Fig. 3.1, which is the chromatographic pattern obtained when the products of a brief digestion are separated on DEAE-cellulose in the presence of 7 *M* urea. Assays of the peaks for alanine acceptor activity indicated that peak III was unchanged alanine tRNA. Peaks I and II were inactive and the position of elution suggested that these peaks were large fragments of the original tRNA. Rechromatography of peaks I and II at elevated temperature provided no evidence for the presence of more than one component in either peak. The oligonucleotide content of each peak was determined after complete digestion with T_1 RNase.

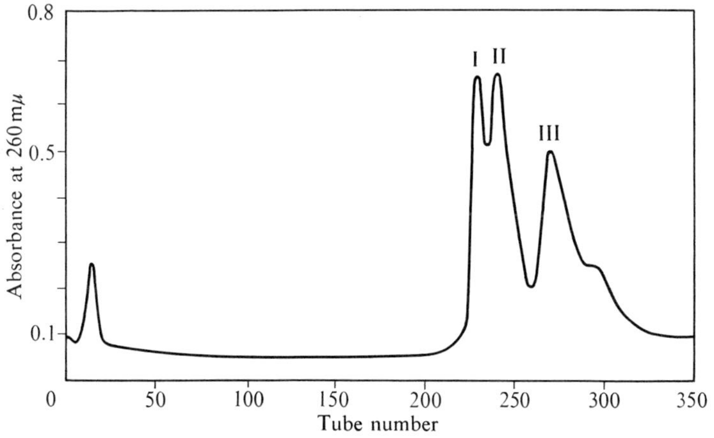

Fig. 3.1. Chromatographic separation of the products from a limited digestion of alanyl tRNA with T_1 RNase (from Penswick and Holley, 1965). Chromatography performed with a DEAE-cellulose column in the presence of 7 *M* urea.

A remarkable example of chromatographic science may be seen in the patterns that were obtained from the separation of the oligonucleotides produced from peaks I and II, shown in Figs. 3.2a and 3.2b. These separations were also performed with DEAE-cellulose in the presence of 7 *M* urea. Under the conditions used, a greater resolution of the peaks is achieved as a result of differences in base composition at the expense of separation strictly according to chain length. The pattern in Fig. 3.2c is from the chromatography of the products obtained from a complete T_1 RNase hydrolysis of alanine tRNA. An analysis of the individual peaks in Figs. 3.2a and 3.2b showed that they were mutually exclusive and complementary, and that the sum of the analyses corresponds to the analysis of the oligonucleotides in the original alanine tRNA in Fig. 3.2c. Furthermore, these analyses permitted the assignment of the oligonucleotides in the left column of Table 3.5 to the left end of alanyl tRNA, and the oligonucleotides in the right column to the right end. The correct positioning of the individual fragments within each half was accomplished as a result of a separate limited digestion with T_1 RNase (Apgar *et al.*, 1965). It was found in this second digestion that 4 of the 9 fragments contained sequences belonging to

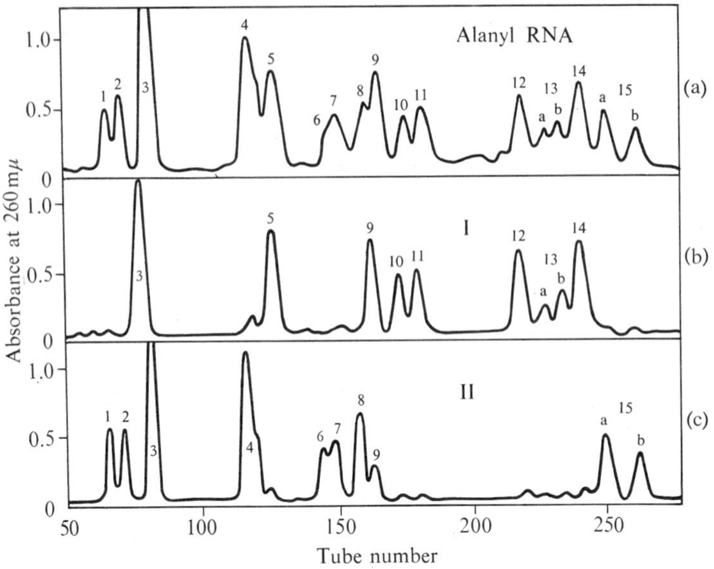

Fig. 3.2. Separation patterns from chromatography of products of further T_1 RNase digestion of peaks I and II of Fig. 3.1. (a) Oligonucleotides from complete T_1 RNase hydrolysis of alanyl tRNA. (b) Oligonucleotides from peak I. (c) Oligonucleotides from peak II.

the left half of the molecule, while the remaining 5 fragments contained oligonucleotides associated with the right-hand end of the molecule. The entire base sequence was finally established by determining the detailed sequences of the oligonucleotides from other partial digests.

DETERMINATION OF BASE SEQUENCE IN PHENYLALANYL TRANSFER RNA

A detailed example of the procedure for the determination of base sequences using the overlap method is provided by the studies of RajBhandary and co-workers (RajBhandary *et al.*, 1968a–d; Chang and RajBhandary, 1968). As indicated above, the first step is a nucleoside composition determination followed by an analysis of the products obtained when yeast phenylalanine tRNA is digested with T_1

Table 3.6. Nucleoside composition of phenylalanine tRNA

16-A	2-hU	1-m^2_2G	1-Gm
15-C	2-Ψ	1-m^2G	1-m^7G
18-G	2-m^5C	1-m′A	1-N
12-U	1-T	1-Cm	

Table 3.7. Oligonucleotides obtained from a pancreatic RNase digest of phenylalanyl tRNA

1. C-OH	8. C-	15. hU	22. G-T-	29. G-G-A-U-
2. C-	9. U-	16. Ψ	23. A-C-	30. A-G-A-A-U-
3. C-	10. U-	17. m^5C	24. A-C-	31. A-G-A-C^m-U-
4. C-	11. U-	18. m^5C	25. G-C-	32. G^m-A-A-N-A-Ψ-
5. C-	12. U-	19. pG-C-	26. A-$\overset{m^2}{G}$-C-	33. G-G-A-G-$\overset{m^7}{G}$-U-
6. C-	13. U-	20. $\overset{m^2_2}{G}$-C-	27. A-G-$\overset{h}{U}$-	34. G-G-G-A-G-A-G-C-
7. C-	14. U-	21. G-U-	28. G-$\overset{m^1}{A}$-U-	

Table 3.8. Oligonucleotides obtained from a T_1 RNase digest of phenylalanine tRNA

1. pG-	8. A-G-	15. C-A-C-C-OH-
2. G-	9. A-G-	16. C-U-C-A-G-
3. G-	10. A-G-	17. A-U-U-U-A-$\overset{m^2}{G}$-
4. G-	11. U-G-	18. A-A-U-U-C-G-
5. G-	12. $\overset{h}{U}$-$\overset{h}{U}$-G-	19. $\overset{m^7}{G}$-U-C-$\overset{m^5}{C}$-U-G-
6. C-G-	13. C-C-A-G-	20. $\overset{m^1}{A}$-U-C-C-A-C-A-G-
7. C-$\overset{m^2_2}{G}$-	14. T-Ψ-C-G-	21. A-C^m-U-G^m-A-A-N-A-Ψ-$\overset{m^5}{C}$-U-G-

and pancreatic RNase. Table 3.6 shows the nucleoside composition of phenylalanine tRNA, Table 3.7 lists the oligonucleotides that are obtained from a pancreatic RNase digestion, and Table 3.8 lists the products of a T_1 digestion. As in the case of yeast alanine tRNA, a comparison of the two lists indicates several overlaps, some of single odd bases and others of sequences of the more common bases. An example of the former case is illustrated below: oligonucleotide 19 from the T_1 RNase list

$$\overset{m^7}{G}\text{-U-C-}\overset{m^5}{C}\text{-U-G-}$$

overlaps with oligonucleotide 33 from the pancreatic RNase list

$$\text{G-G-A-G-}\overset{m^7}{\text{G}}\text{-U-}$$

because there is only one m^7G in yeast phenylalanine tRNA. The combined sequence should therefore be

$$\overset{33}{\overline{\text{G-G-A-G-}\overset{m^7}{\text{G}}\text{-U}}}\text{-C-}\overset{m^5}{\text{C}}\text{-U-G-}$$

(underline, 19: $\text{G-U-C-}\overset{m^5}{\text{C}}\text{-U-G-}$)

where the overline indicates the oligomer from the pancreatic digest and the underline indicates the oligomer from the T_1 digest.

An example of the second case is the overlap between pancreatic oligomer 30 and T_1 oligomer 18. An examination of Tables 3.7 and 3.8 shows that the sequence of A-A-U- occurs only in these oligomers and that they can be fitted together as

$$\overset{30}{\overline{\text{A-G-A-A-U}}}\text{-U-C-G-}$$

(underline, 18: A-A-U-U-C-G-)

A list of all the overlapping sequences from the complete digestion of phenylalanine tRNA is given in Table 3.9. Also indicated there is the base which precedes the overlapping sequence. A slightly different listing of the overlapping sequences could be made to account precisely for all of the nucleotide residues, as in Table 3.5. The list would include 13 separate fragments in addition to the right and left termini and would indicate the large number of possible ways the overlapping sequences could be fitted together. The type of listing shown in Table 3.9 is more useful, however.

SEQUENCES FROM PARTIAL HYDROLYSIS

The next step, as before, is to obtain additional areas of overlapping sequences by carrying out partial digestions with T_1 and pancreatic RNase. When phenylalanyl tRNA is hydrolyzed briefly with T_1 RNase, several large oligonucleotides can be isolated by a series of

Table 3.9. Summary of known overlapping sequences of phenylalanine tRNA

No.	Fragments above	Modifications	Sequence	Fragments below
1.	19		pG-C-	1
2.	33	m^7, m^5	Y-G-G-A-G-G-U-C-C-U-G-	19
3.	31, 32, 17, 9	m^5	Y-A-G-A-C^m-U-G^m-A-A-Y-A-Ψ-C-U-G-	21
4.	28, 2, 3, 23	m^1	Y-G-A-U-C-C-A-C-A-G-	20
5.	30, 10, 4		Y-A-G-A-A-U-U-C-G-	18
6.	29, 11, 12, 26	m^2	Y-G-G-A-U-U-U-A-G-C-	2, 17
7.	3, 4		Y-G-G-G-A-G-A-G-C-	3, 4, 8, 9
8.	27, 15	h h	Y-A-G-U-U-G-	12
9.	13, 5		R-C-U-C-A-G-	16
10.	24, 1		R-C-A-C-C-OH-	15
11.	22, 16, 6		Y-G-T-Ψ-C-G-	14
12.	7		R-C-C-A-G	13
13.	20	m_2^2	R-C-G-C-	7

chromatographic fractionations. The data obtained from the analysis of these larger fragments permitted the reconstruction of phenylalanyl tRNA as follows:

Of all the large oligonucleotides obtained from partial T_1 RNase digestion, there were seven of special interest. These were designated by the letters *a*, *b*, *c*, *d*, *g*, *h*, and *j*. A partial digestion of phenylalanyl tRNA with pancreatic RNase yielded, among other oligonucleotides, a fragment designated *f*.

Further digestion of fragment *c* with T_1 RNase produced two oligonucleotides:

1. C-A-C-C-OH
2. A-A-U-U-C-G-

Oligomer 1 is a tetranucleoside triphosphate and must be the right-hand terminus of phenylalanyl tRNA. Oligomer 2 must be to the left of 1 to yield the end sequence.

A-A-U-U-C-G-C-A-C-C-OH

Oligonucleotide 5 of Table 3.9 indicates that this end sequence may be expanded to

Y-A-G-A-A-U-U-C-G-C-A-C-C-OH

Fragment *g*, when digested further with T_1 RNase, yielded the oligonucleotides

1. C-A-C-C-OH
2. A-A-U-U-C-G-
3. $\overset{m^1}{A}$-U-C-C-A-C-A-G-

As oligomers 1 and 2 are part of the right-hand end sequence, oligomer 3 must be to the left to yield

$\overset{m^1}{A}$-U-C-C-A-C-A-G-A-A-U-U-C-G-C-A-C-C-OH

Oligomer 4 from Table 3.9 indicates that a Y-G- is to the left of fragment *g*.

Fragment *j*, when digested with T_1 RNase, was found to contain all of the oligonucleotides of fragment *g* plus

4. U-G-
5. T-Ψ-C-G-
6. $\overset{m^7}{G}$-U-C-$\overset{m^5}{C}$-U-G-

Oligonucleotide 2 of Table 3.9 is

Y-G-G-A-$\overset{m^7}{G}$-G-U-C-$\overset{m^5}{C}$-U-G-

and contains oligomer 6 of *j*. There is no A-G- dimer in T_1 digest of *j*, an indication that there are no nucleotides to the left of oligomer 6 and that in fact oligomer 6 must be the left end of *j*. This means that oligonucleotides 4 and 5 are to the right of 6 and to the left of fragment *g*:

m7 m5 m1
G-U-C-C-U-G-(U-G-,T-Ψ-C-G-)-A-U-C-C-A-C-A-G-A-A-
U-U-C-G-C-A-C-C-OH

Additional information was needed to determine the correct sequence of the oligomers inside the parentheses. It was provided by a fragment *f*, obtained from a partial pancreatic RNase digestion of phenylalanine tRNA. Fragment *f*, when further hydrolyzed with T_1 RNase, produced

1. U-
2. G-
3. G-
4. A-G-
5. m5 m5
G-U-C-C-U-G-

The mononucleotide U- must be the right-hand end of *f* because U- is the only product of a T_1 RNase digestion of *f* that does not end in G-. The remaining products of this digestion are all accounted for in the combined oligonucleotide 2 of Table 3.9, which with U- at the right end becomes

m7 m5
Y-G-G-A-G-G-U-C-C-U-G-U-

This indicates that the correct order of oligomers within the parentheses of *j* is U-G- at the left and T-Ψ-C-G- at the right. The sequence of fragment *j* is therefore

m7 m5 m1
G-U-C-C-U-G-U-G-T-Ψ-C-G A-U-C-C-A-C-A-G-A-A
U-U-C-G-C-A-C-C-OH

Considering the structure of *j* and oligonucleotide 2 from Table 3.9, the sequence of 35 nucleosides at the right-hand end of phenylalanine tRNA may be written

m7 m5 m1
Y-G-G-A-G-G-U-C-C-U-G-U-G-T-Ψ-C-G-A-U-C-C-A-C-A
G-A-A-U-U-C-G-C-A-C-C-OH

When fragment *a* was hydrolyzed completely with T_1 RNase, the products were

1. G-
2. pG-
3. C-G-
4. A-U-U-U-A-$\overset{m^2}{\text{G}}$- (plus degradation products of this oligomer)

Oligomer 2 is obviously the nucleotide at the left terminus of the tRNA. In Table 3.9, the left terminus oligonucleotide 1 is pG-C-. The only oligonucleotide in fragment *a* that contains C is oligomer 3, C-G-. The left end sequence must therefore be pG-C-G-. The question of which of the two products, G- or oligomer 4, follows the end sequence is resolved by consideration of sequence 6 of Table 3.9. This sequence is

$$\text{Y-G-G-A-U-U-U-A-}\overset{m^2}{\text{G}}\text{-C-}$$

and it contains oligomer 4 from fragment *a*. The m^2G- is followed by a C- and not a G-. Since the remaining G- of fragment *a* cannot follow oligomer 4 of fragment *a*, it must precede oligomer 4, and the sequence of *a* is

$$\text{pG-C-G-G-A-U-U-U-A-}\overset{m^2}{\text{G}}\text{-}$$

The left-hand end is

$$\text{pG-C-G-G-A-U-U-U-A-}\overset{m^2}{\text{G}}\text{-C-}$$

Fragment *d* was first treated with phosphomonoesterase and then hydrolyzed with T_1 RNase. This yielded all of the components of fragment *a* and, in addition, a pentanucleoside tetraphosphate with the sequence C-U-C-A-G-. This oligonucleotide must be to the right of *a* because *a*, as shown above, is the left terminus. The sequence at this end is therefore

$$\text{pG-C-G-G-A-U-U-U-A-}\overset{m^2}{\text{G}}\text{-C-U-C-A-G-}$$

Fragment *b*, when treated with T_1 RNase, yielded essentially the following oligonucleotides:

1. G-
2. G-
3. A-G-
4. A-G-
5. $\overset{h}{U}\text{-}\overset{h}{U}\text{-G-}$
6. $\text{C-}\overset{m^2}{G}\text{-C-C-A-G-}$
7. $\text{A-}C^m\text{-U-}G^m\text{-A-A-N}'\text{-A-}\Psi\text{-}\overset{m^5}{C}\text{-U-G-}$

Oligomer 7 contained an unknown nucleoside N that was converted into a fluorescent form N′ during chromatographic manipulations. Later, this was shown to be isopentyladenosine, iA. Oligomer 7 could be identified from its products after a pancreatic RNase digestion. The products showed that oligomer 7 of fragments *b* was a part of oligonucleotide 3 of Table 3.9:

$$\text{Y-A-G-A-}C^m\text{-U-}G^m\text{-A-A-N-A-}\Psi\text{-}\overset{m^5}{C}\text{-U-G-}$$

and that, in tRNA, oligomer 7 was preceded by a Y-A-G- sequence. This finding ruled out the possibility that oligomers 3 and 4 of fragment *b* preceded oligomer 7 and indicated instead that oligomer 6, with an end sequence C-A-G-, was the oligonucleotide to the left of 7. This combined sequence is

$$\text{C-}\overset{m_2^2}{G}\text{-C-C-A-G-A-}C^m\text{-U-}G^m\text{-A-A-N-A-}\Psi\text{-}\overset{m^5}{C}\text{-U-G-}$$

This leaves two G-, two A-G-, and one $\overset{h}{U}\text{-}\overset{h}{U}\text{-G-}$ in fragment *b* unaccounted for. A pancreatic RNase digest of fragment *b* yielded, among other nucleotides, the oligomer G-G-G-A-G-A-G-C-. This sequence accounts for the two G- and the two A-G-. In addition, no $\text{A-G-}\overset{h}{U}\text{-}$ was found, as would be required by oligonucleotide 8 of Table 3.9. These facts can be reconciled only if oligomer 5, $\overset{h}{U}\text{-}\overset{h}{U}\text{-G-}$, is the left end of

Table 3.10. Summary of sequence analysis of tRNAs

tRNA	Source	Complete Analysis	Isolation and Purification
1. Alanine	Yeast	Holley *et al.* (1965)	Countercurrent distribution (CCD) with phosphate buffer-formamide isopropanol system (Holley *et al.*, 1963)
2. Serine I, II	Yeast	Zachau *et al.* (1966)	CCD in tri-*n*-butylamine solvent (Zachau *et al.*, 1961); see also Doctor, 1968
3. Tyrosine	Yeast	Madison *et al.* (1966)	Same as No. 1
4. Valine	Yeast	Baev *et al.* (1967a, b)	CCD (Apgar *et al.*, 1962); isolation of valyl tRNA from a column of polyacrylic acid hydrazide-agar gel, after periodate oxidation of "uncharged"tRNAS (Knorre *et al.*, 1968)
5. Phenylalanine	Yeast	RajBhandary and Chang (1968a)	CCD (Hoskinson and Khorana, 1965; RajBhandary *et al.*, 1968a–d)
6. Valine	*T. utilis*	Mizutani *et al.* (1968)	Column chromatography on DEAE Sephadex (Miyazaki *et al.*, 1966; Miyazaki and Takemura, 1966; Miyazaki *et al.*, 1967
7. Isoleucine	*T. utilis*	Takemura *et al.* (1969)	Same as No. 6 above
8. Serine	Rat	Staehelin *et al.* (1968)	Partition chromatography (Muench and Berg, 1966); chromatography on methylated albumin-kieselguhr columns (Sueoka and Yamane, 1962; Melchers and Zachau, 1965
9. Phenylalanine	Wheat	Dudock *et al.* (1968)	—
10. Tyrosine suppressor+ III	*E. coli* ϕ80 bacteriophage	Goodman *et al.* (1968)	^{32}P-labeled tRNA; column chromatography on DEAE-Sephadex (Cherayil and Bock, 1965); separation of charged tyrosyl tRNA from periodate-oxidized impurities on polyacrylic acid hydrazide column (vonPortatius *et al.*, 1961); solubility properties of poly-β-benzyl-L-aspartyl tyrosine tRNA (Katchalsky *et al.*, 1966)
11. Methionine	*E. coli*	Cory *et al.* (1968)	^{32}P-tRNA; column chromatography on DEAE-Sephadex (Nishimura *et al.*, 1967); column chromatography on benzoylated DEAE cellulose (Gillam *et al.*, 1967)
12. N-formyl-methionine	*E. coli*	Dube *et al.* (1968, 1970)	Same as No. 11
13. Valine	*E. coli*	Yaniv and Barrell (1969)	Same as No. 11
		Harada *et al.* (1969)	Same as No 11
14. Tyrosine	*T. utilis*	Hashimoto *et al.* (1969)	Same as No. 11
15. Phenylalanine	*E. coli*	Barrell and Sanger (1969)	^{32}P-tRNA isolated either by specific binding with poly U and ribososomes or by reversed phase partition chromatography (Kelmers *et al.*, 1965)

Table 3.10. (continued)

Modified Bases	Separation and Analysis of Oligonucleotides
Ψ-2,hU-2,rT,I,U*,m^{1}G,m^{1}I,m$^2_2$G	Column chromatography on DEAE-cellulose, DEAE-Sephadex; two-dimensional paper fingerprint: electrophoresis in first dimension followed by chromatography in second dimension
Ψ-3,hU-3,rT,I,Um,Gm,m^{5}C,acC,m$^2_2$G,iA	Similar to No. 1
Ψ-3,hU-6,rT,m^{1}A,m^{5}C,m^{2}G,Gm,m$^2_2$G,iA	Same as No. 1
Ψ-4,hU-4,rT,I,m^{1}A,m^{5}C,m^{1}G	Similar to No. 1; two-dimensional paper fingerprint according to Krutilina *et al.* (1965), Li *et al.* (1966)
Ψ-2,hU-2,rT,m^{1}A,m^{5}C-2,m^{2}G,m^{7}G,Cm,Gm,m$^2_2$G,iA	Similar to No. 1
Ψ-4,hU-3,rT,I,m^{1}A,m^{5}C,m^{1}G	Column chromatography on DEAE-cellulose and AGI
	Same as No. 6 above
Ψ-2,hU-3,Ψm,rT,I,m^{1}A,m^{3}C-2,m^{5}C,acC,Gm,Um,m$^2_2$G,iA	Column chromatography; thin-layer chromatography
Ψ-2,rT,4S-2,A*,Gm	Two-dimensional fingerprint method (see Chapter 4): electrophoresis on cellulose acetate followed by transfer to DEAE-paper and electrophoresis in second dimension and radioautography; Sanger *et al.* (1965)
Ψ-2,hU-3,rT,X,U*,A*,C+,Gm,m^{7}G	Two-dimensional fingerprint as in No. 10; homochromatography; Brownlee *et al.* (1968)
Ψ,hU,rT,U*,m^{7}G,Cm	Same as No. 11
Ψ,hU,rT,m*A,m^{7}G,^{4}S,X	Same as No. 11
Same	Same as No. 8
Ψ-4,hU-6,rT,m^{1}A,iA,m^{5}C,m^{1}G m^{2}G,m$^2_2$G,Gm	Similar to No. 6
Ψ-3,hU-2,rT,siA,m^{7}G,^{4}S,X	Same as No. 11

fragment *b*, followed by oligomers 1, 2, 3, 4, 6, and 7 to yield the sequence

```
h h                 m2/2
U-U-G-G-G-A-G-A-G-C-G-C-C-A-G-A-C-U-G-A-A-N-
  m5
A-Ψ-G-U-G-
```

When fragment *h* was hydrolyzed completely with T_1 RNase, all of the oligonucleotides of fragment *b* were obtained plus the oligonucleotide C-U-C-A-G-. The question was whether C-U-C-A-G- was to the right or left of *b*. It had already been shown that there is only one R-C-U-C-A-G- sequence in phenylalanine tRNA (Table 3.8), and it had been assigned to the left terminal fragment *d*. Obviously, *d* can only be followed by *h*, so that C-U-C-A-G- must precede the *b* portion of fragment *h*. The combined left end sequence then becomes

```
                  m2              h h
pG-C-G-G-A-U-U-U-A-G-C-U-C-A-G-U-U-G-G-G-A-G-A-
    m2/2                         m5
G-C-G-C-C-A-G-A-C-U-G-A-A-N-A-Ψ-C-U-G-
```

This left-hand portion of phenylalanine tRNA contains 42 nucleoside residues. The right-hand portion was seen previously to contain 35 nucleoside residues. The last two residues of the left-hand portion, U-G-, overlap with the first two residues of the right-hand portion, Y-G-, to form the 75-nucleoside sequence of phenylalanine tRNA:

```
         m2      h h         m2/2
pGCGGAUUUAGCUCAGUUGGGAGAGCGCCAGA-
         m5     m7  m5           m1
CᵐUGᵐAANAΨCUGGAGGUCCUGUGTΨCGAUCCA-
CAGAAUUCGCACC-OH
```

A possible secondary structure for this and other tRNAs is given in Chapter 12.

PRIMARY STRUCTURE OF OTHER tRNAs

With an increasing number of investigators interested in nucleic acid sequence analysis, it is natural that certain variations and improvements in the successful methods have been developed. Two of the more significant developments have been the use of column chromatography for the isolation and purification of individual tRNAs, and a two-dimensional electrophoresis system for the identification and determination of oligonucleotides from RNase digests of radioactive RNA. A summary of the application of these newer methods as well as the older ones in the sequence analysis of tRNA appears in Table 3.10. The two-dimensional electrophoresis procedure was initially successful in establishing the primary structure of 5S RNA; this is discussed in greater detail in Chapter 4. Table 3.10 also lists the modified bases found in each of the tRNAs in order to provide an idea of the number of overlaps found on this basis.

It is clear that much of the most recent work has been carried out with radioactive tRNA. As a result, sequence analysis could be performed on much smaller amounts of material. This, together with fingerprint procedures for the rapid identification and determination of oligonucleotides, appears to be the choice for future work in this field.

REFERENCES

Apgar, J., Everett, G. A., and Holley, R. W. (1965). *Proc. Natl. Acad. Sci.*, **53**, 546.

Apgar, J., Holley, R. W., and Merrill, S. H. (1962). *J. Biol. Chem.*, **237**, 769.

Baev, A. A., Venkstern, T. V., Mirzabekov, A. D., Krutilina, A. I., Axelrod, V. D., Li, L., Fodor, I., Kasarinova, L. Y., Engelhardt, V. A. (1967a). *Structure and Function of Transfer RNA and 5S-RNA*, L. O. Froholm and S. G. Laland (eds.), 1968. Academic Press, London.

Baev, A. A., Venkstern, T. V., Mirzabekov, A. D., Krutilina, A. I., Li, L., and Axelrod, V. D. (1967b). *Mol. Biol.*, **1**, 754.

Barrell, B. G., and Sanger, F. (1969). *FEBS Letters*, **3**, 275.

Brownlee, G. G., Sanger, F., and Barrell, B. G. (1968). *J. Mol. Biol.*, **34**, 379.

Chang, S. H., and RajBhandary, U. L. (1968). *J. Biol. Chem.*, **243**, 598.

Cherayil, J. D., and Bock, R. M. (1965). *Biochemistry*, **4**, 1174.

Cory, S., Marcker, K. A., Dube, S. K., and Clark, B. F. C. (1968). *Nature*, **220**, 1039.

Doctor, B. P. (1968). In *Methods in Enzymology*, Volume 12, part B, L. Grossman and K. Moldave (eds.). Academic Press, New York.

Dube, S. K., Marcker, K. A., Clark, B. F. C., and Cory, S. (1968). *Nature*, **218**, 232.

Dube, S. K., Marcker, K. A., Clark, B. F. C., and Cory, S. (1970). *FEBS Letters*, **9**, 168.

Dudock, B. S., Katz, G., Taylor, E. K., and Holley, R. W. (1969). *Proc. Natl. Acad. Sci.*, **62**, 941.

Gillam, I., Millward, S., Blew, D., von Tigerstrom, M., Wimmer, E., and Tener, G. M. (1967). *Biochemistry*, **6**, 3043.

Goodman, H. M., Abelson, J., Landy, A., Brenner, S., and Smith, J. D. (1968). *Nature*, **217**, 1019.

Hall, R. H. (1965). *Biochemistry*, **4**, 661.

Hall, R. H. (1971). *The Modified Nucleosides in Nucleic Acids*, Columbia University Press, New York.
Harada, F., Kimura, F., and Nishimura, S. (1969). *Biochim. Biophys. Acta*, **195**, 590.
Hashimoto, S., Miyazaki, M., and Takemura, S. (1969). *J. Biochem.* (Japan), **65**, 659.
Holley, R. W., Apgar, J., Everett, G. A., Madison, J. T., Merrill, S. H., and Zamir, A. (1963). *Cold Spring Harbor Symp. Quant. Biol.*, **28**, 117.
Holley, R. W., Everett, G. A., Madison, J. T., and Zamir, A. (1965). *J. Biol. Chem.*, **240**, 2122.
Hoskinson, R. M., and Khorana, H. G. (1965). *J. Biol. Chem.*, **240**, 2129.
Katchalsky, E., Yanofsky, S., Novogrodsky, A., Galenter, Y., and Littauer, U. Z. (1966). *Biochim. Biophys. Acta*, **123**, 641.
Kelmers, A. D., Novelli, G. D., and Stulberg, M. P. (1965). *J. Biol. Chem.*, **240**, 3979.
Knorre, D. G., Misina, S. D., and Sandakhchiev, L. S. (1968). *Izv. Sibirsk Otd. Akad. Nauk SSSR*, N11, 143.
Krutilina, A. I., Mirzabekov, A. D., Venkstern, T. V., and Baev, A. A. (1965). *Biokhimia*, **30**, 1225.
Li, L., Venkstern, T. V., Mirzabekov, A. D., Krutilina, A. I., and Baev, A. A. (1966). *Biokhimia*, **31**, 117.
Madison, J. T., Everett, G. A., and Kung, H. (1966). *Science*, **153**, 531.
Melchers, F., and Zachau, H. G. (1965). *Biochim. Biophys. Acta*, **95**, 380.
Miyazaki, M., and Takemura, S. (1966). *J. Biochem. (Japan)*, **60**, 526.
Miyazaki, M., Karata, M., and Takemura, S. (1966). *J. Biochem. (Japan)*, **60**, 519.
Miyazaki, M., Karata, M., Nakagawa, K., and Takemura, S. (1967). *J. Biochem. (Japan)*, **62**, 161.
Mizutani, T., Miyazaki, M., and Takemura, S. (1968). *J. Biochem.* (Japan), **64**, 839.
Muench, K. H., and Berg, P. (1966). *Biochemistry*, **5**, 970.
Nishimura, S., Harada, F., Narushima, U., and Seno, T. (1967). *Biochim. Biophys. Acta*, **142**, 133.
Penswick, J. R., and Holley, R. W. (1965). *Proc. Natl. Acad. Sci.*, **53**, 543.
RajBhandary, U. L., and Chang, S. H. (1968a). *J. Biol. Chem.*, **243**, 592.
RajBhandary, U. L., Faulkner, R. D., and Stuart, A. (1968b). *J. Biol. Chem.*, **243**, 575.
RajBhandary, U. L., Stuart, A., Hoskinson, R. M., and Khorana, H. G. (1968c). *J. Biol. Chem.*, **243**, 565.
RajBhandary, U. L., Stuart, A., and Chang, S. H. (1968d). *J. Biol. Chem.*, **243**, 584.
Sanger, F., Brownlee, G. G., and Barrell, B. G. (1965). *J. Mol. Biol.*, **13**, 373.
Sanger, F., and Tuppy, H. (1951). *Biochem. J.*, **49**, 463, 481.
Sober, H. A. (ed.) (1968). *Handbook of Biochemistry*. The Chemical Rubber Co., Cleveland, Ohio.
Staehelin, M., Rogg, H., Baguley, B. C., Gensberg, T., and Wehrli, W. (1968). *Nature*, **219**, 1363.
Sueoka, N., and Yamane, T. (1962). *Proc. Natl. Acad. Sci.*, **48**, 380.
Takemura, S., Murakami, M., and Miyazaki, M. (1969). *J. Biochem.*, **65**, 553.
von Portatius, H., Doty, P., and Stephenson, M. (1961). *J. Am. Chem. Soc.*, **83**, 3351.
Yaniv, M., and Barrell, B. G. (1969). *Nature*, **222**, 278.
Zachau, H. G., Tada, M., Lawson, W. B., and Schweiger, M. (1961). *Biochim. Biophys. Acta*, **53**, 221.
Zachau, H. G., Dutting, D., and Feldmann, H. (1966). *Z. Physiol. Chemie*, **347**, 212.

4.

THE OVERLAP METHOD AND 5S RNA

The primary structures of *E. coli* MRE600 5S RNA (Brownlee *et al.*, 1967) and KB carcinoma cell 5S RNA (Forget and Weissman, 1967) have been determined by the overlap method used in tRNA analysis. However, the details of the application of this method to the base sequence analysis of 5S RNA differ significantly from those in tRNA analysis. These differences are due mainly to the availability of source material and the structural features of 5S RNA and are of enough general significance to warrant a separate discussion.

The 5S RNA was first described by Rosset and Monier (1963) and Rosset *et al.* (1964) and appears to be an integral structural part of all ribosomes. It is RNA of comparatively low molecular weight. It contains approximately 120 nucleotides and, in contrast to tRNA, no "minor" bases. The composition of *E. coli* MRE600 5S RNA is $A_{23}C_{36}G_{41}U_{20}$, and of KB cell 5S RNA is $A_{22}C_{33}G_{39}U_{26}$. The relatively low A content and the relatively high G content of both 5S RNAs lessens the likelihood of finding overlapping sequences in T_1 RNase and pancreatic RNase digests. Moreover, the absence of the "minor" bases provides no opportunity for finding overlapping sequences based on the occurrence of a unique base. As a result, it might be expected that fewer instances of overlapping sequences would be found in T_1 and pancreatic RNase digests of 5S RNA than were found in tRNA. That

Table 4.1. Sequence and molar yield of oligonucleotides derived from pancreatic and T_1 RNase hydrolysis of E. coli MRE600 5S RNA

Pancreatic RNase Oligonucleotide[a]	Moles	T_1 RNase	Moles
a1 C	19	t1 G	11
a2 U	6	t2 C-G	5
a3 A-C	1	t3 A-G	2
a4 G-C	7	t4 U-G	4
a5 A-U	2	t5 C-C-G	4
a6 G-A-C	1	t6 A-A-G	4
a7 A-G-C	2	t7 C-C-A-G	1
a8 G-A-A-C	1	t8 A-A-A-C-G	1
a9 G-A-A-A-C	1	t9 C-C-U-G	1
a10 G-G-C	2	t10 U-A-G	4
a11 A-G-G-C	1	t11 A-U-G	1
a12 G-U	3	t12 A-A-C-U-G	1
a13 G-A-U	1	t13 A-A-C-U-C-A-G	1
a14 A-G-U	1	t14 A-C-C-C-C-A-U-G	1
a15 G-G-U	3	t15 U-C-C-C-A-C-C-U-C	1
a16 A-G-G-G-A-A-C	1	t16 U-C-U-C-C-C-A-U-G	1
a17 A-G-A-A-G-U	1	t17 C-C-U-U-A-G	[b]
a18 G-A-G-A-G-U	1	t18 pU-G	1
a19 G-G-G-G-U	1	t19 C-A-U-OH	1
a20 pU	1		
a21 A-U-OH	1		

[a] Oligonucleotides are shown without terminal phosphates and are from only one of two species of 5S RNA analyzed.

[b] Identified in strain MRE600 and no other strains studied, including CA 265, on which much of the subsequent work was carried out.

this is true can be seen in the lists of products obtained from the T_1 and pancreatic RNase hydrolysis of *E. coli* MRE600 5S RNA (Table 4.1) and KB cell 5S RNA (Table 4.2). The oligonucleotides were separated and their sequences established essentially by the methods of Sanger *et al.* (1965) for the separation and identification of ^{32}P-labeled oligonucleotides by two-dimensional electrophoresis. The identification of many oligonucleotides can be made solely (and rapidly) on the basis of their positions in the two-dimensional pattern or fingerprint. Furthermore, because the final estimation of each oligonucleotide depends on its radioactivity, a comparatively large number of experiments may be

Table 4.2. Sequence and molar yield of oligonucleotides derived from pancreatic and T_1 RNase hydrolysis of KB cell 5S RNA

Pancreatic RNase Oligonucleotide	Moles	T_1 RNase	Moles
a1 C-	17	t1 G-	15
a2 U-	13*	t2 C-G-	1
a3 A-C-	5	t3 A-G-	1
a4 A-U-	1	t4 U-G-	1
a5 G-C-	4	t5 C-A-G-	1
a6 G-U-	2	t6 A-A-G-	1
a7 A-G-U-	1	t7 U-C-G-	1
a8 G-A-U-	2	t8 C-U-G-	1
a9 G-G-C-	1	t9 U-A-G-	1
a10 A-A-G-C-	1	t10 A-U-G-	1
a11 A-A-G-C-	1	t11 C-C-C-G-	1
a12 G-A-A-C-	1	t12 A-C-C-G-	1
a13 A-G-G-C-	1	t13 A-A-C-G-	1
a14 G-G-A-U-	1	t14 C-C-U-G-	2
a15 G-G-G-C-	1	t15 U-C-U-G-	1
a16 G-G-G-U-	1	t16 U-U-A-G-	1
a17 A-G-G-G-U-	1	t17 C-U-A-A-G-	1
a18 G-G-A-A-G-C-	1	t18 U-A-C-U-U-G-	1
a19 G-G-G-A-A-U-	1	t19 A-U-C-U-C-G-	2
a20 G-G-G-A-G-A-C-	1	t20 U-C-U-A-C-G-	1
a21 pG-U-	1	t21 A-A-U-A-C-C-G-	1
		t22 C-C-A-U-A-C-C-A-C-C-C-U-G-	1
		t23 pG-	1
		t24 C-U-U-OH-	1*

* Evidence was obtained for the presence of an equal amount of a second species of 5S RNA with a right-hand terminal sequence, C-U-U-U-OH, and consequently one more U- in the pancreatic RNase digest.

carried out on a small amount of 5S RNA. This experimental method, including some special features, is described in greater detail in Chapter 11.

The equivalent amounts of nucleotides and oligonucleotides shown in Tables 4.1 and 4.2 represent yields corrected to the actual amounts in the 5S RNAs. In most instances, the experimental values were very close to the theoretical values. In the *E. coli* 5S RNA, there were no significant overlapping sequences; that is, no oligonucleotide from the pancreatic RNase digest could be used to join any two oligonucleotides from the T_1 RNase digest. In addition, the incidence of matching sequences was also rare. Oligonucleotide a17 from the pancreatic list is seen to include one of the oligonucleotides, t6, from the T_1 list:

a17
A-G-A-A-G-U
t6

Pancreatic oligonucleotide a9 matches with T_1 oligonucleotide t8:

a9
G-A-A-A-C-G
t8

Beyond these two sequences, there are no examples of a unique fit.

With KB cell 5S RNA, the situation is only slightly better. Oligonucleotides from the pancreatic RNase list (upper) may be fitted with oligonucleotides from the T_1 list (lower) in the following ways:

a19 a3 a1
G-G-G-A-A-U-A-C-C-G-
t1 t1 t21

a18
G-G-A-A-G-C-
t1 t6

a20
G-G-G-A-G-A-C-C-G-
t12

a12
G-A-A-C-G-
t13

a2 a11
C-U-A-A-G-C-
t17

It is apparent that these overlaps provide only a small amount of help in the determination of base sequences in 5S RNA. The complete digest lists do reveal the end sequences to be:

E. coli—left, pU-G-(t18); right, C-A-U-OH(t19)
KB cell—left, pG-U-(a21); right, C-U-U-OH(t24)

and they serve as a catalog of sequences to be used in further work with partial hydrolysis products.

LONG-CHAIN FRAGMENTS FROM KB CELL 5S RNA

The methods used to obtain partial hydrolysis products from KB cell 5S RNA were different from those used with *E. coli* 5S RNA. Forget and Weissman elected to work mainly with T_1 RNase partial digestions of KB cell 5S RNA. They found that, in order to arrange all of the oligonucleotides into one linear sequence, more than 35 fragments from partial digestions needed to be analyzed. The long-chain oligonucleotides obtained from these partial hydrolyses with T_1 nuclease were first separated on columns (200 × 0.6 cm) of microgranular DEAE-cellulose with gradients of NaCl in buffered (pH 7.5) 7 *M* urea. The individual peaks eluted from this column were then chromatographed either on a 200 × 0.45 cm column of DEAE-Sephadex, NaCl gradient in 7 *M* urea, pH 3.5, or on a heated column (55°) of DEAE-cellulose (200 × 0.6 cm) eluted with a NaCl gradient in 7 *M* urea buffered at pH 7.5.

In most instances, the fragments isolated by this procedure proved to be homogeneous segments of the original molecule, containing 1 mole, or multiples of 1 mole, of specific oligonucleotides. The oligonucleotide composition of each fragment was determined by complete digestion with T_1 RNase or pancreatic RNase followed by fractionation of the products by the electrophoresis procedure. The pattern obtained in a "fingerprint" was compared to a similar fractionation of whole 5S RNA digest to identify the oligonucleotides present. Each oligonucleotide from the partial fragment was assigned the sequence of the oligonucleotide occurring in the same position in the pattern of fractionation of the whole 5S RNA digest. In doubtful cases, alkaline hydrolysis confirmed the identification of an oligonucleotide.

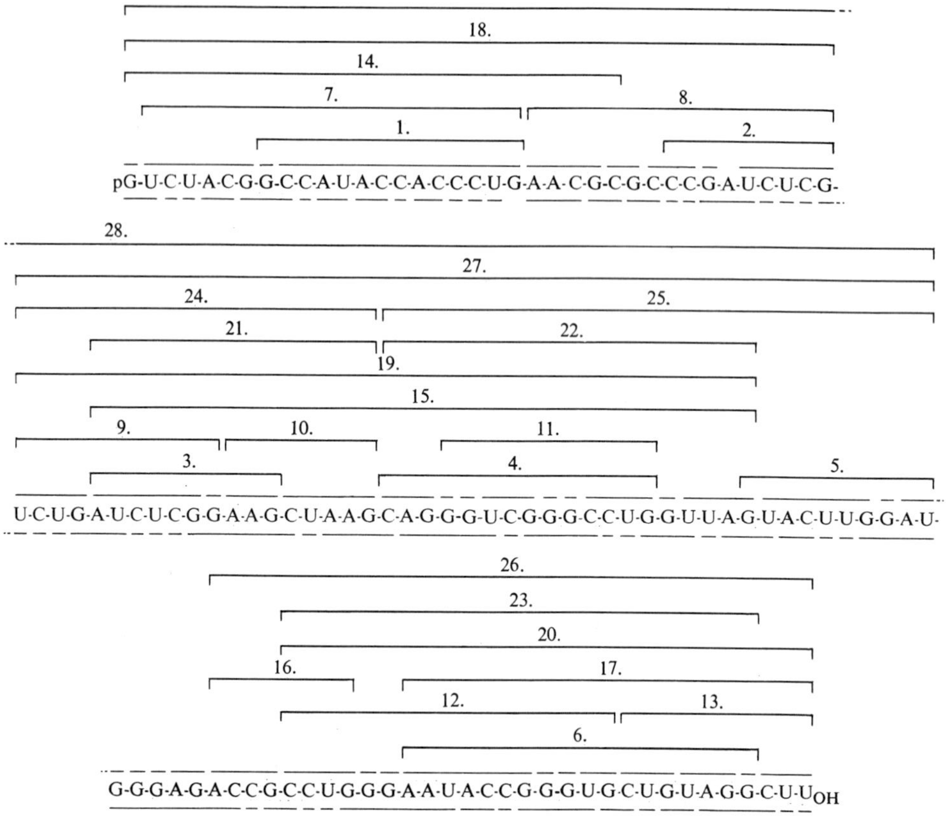

Fig. 4.1. Diagrammatic summary of the sequence analysis of KB carcinoma cell 5S RNA. The bracketed portions illustrate the overlapping sequences obtained from the products of a limited digestion with pancreatic or T_1 RNases (from Forget and Weissmann, 1967).

A diagrammatic summary of the results is shown in Fig. 4.1. All of the oligonucleotides listed in Table 4.2 can be grouped into three general regions of the molecule. The order of the oligonucleotides in the smaller fragments was determined by complete digestion with T_1 and pancreatic RNase. Longer fragments, overlapping these shorter fragments of known sequence, permitted a lengthening of the known sequence until finally all of the oligonucleotides were placed in correct order within each of the three regions.

Once the sequence of each of the three regions was established, it was possible to arrange them in the proper order: one region comprised the left end, one was the right end, and one was the middle. The precise overlap between the left end and the middle was obtained from one other very large fragment, while that between the middle and the right end was provided by oligomer a20 in Table 4.2. This oligomer, G-G-G-A-G-A-C-, was the only oligonucleotide not accounted for in all of the T_1 RNase partial digestion products. As pointed out above, it is the only pancreatic RNase product ending in G-A-C- and must therefore overlap with the oligonucleotide t12 A-C-C-G- to yield

a20
G-G-G-A-G-A-C-C-G-
t12

The oligonucleotide A-C-C-G- is the only T_1 RNase product that begins with A-C-, and this sequence is present at the left end of the fragment containing the right-hand terminus of the 5S RNA, thus providing the correct overlapping sequence between the middle fragment and the right end fragment.

LONG-CHAIN FRAGMENTS FROM E. coli 5S RNA

Sanger and co-workers, on the other hand, chose to augment the information from partial T_1 and pancreatic digests through the use of other nucleases as well as chemical modification of the 5S RNA. The enzyme that was found to be most useful was the acid ribonuclease from spleen. Although this enzyme has no absolute specificity, it shows preference for splitting at adenosine residues and thus gives somewhat different products from those obtained from T_1 and pancreatic ribonucleases. The 5S RNA was modified in two ways: treatment with 1-cyclohexyl-3-(2 morpholinyl-(4)-ethyl)carbodiimide-metho-*p*-toluenesulfonate (CMCT) to yield the water-soluble carbodiimide derivatives, *in situ*, of uridine and guanosine, but not of cytidine, so that specific splitting of the RNA chain at the cytidine residues could be achieved with pancreatic RNase; and methylation of guanosine residues preferentially with dimethyl sulfate to render them resistant to T_1 RNase hydrolysis.

Attempts to fractionate the longer oligonucleotides from the partial T_1 and pancreatic RNase digestion were made using 7% formic acid for the DEAE-paper dimension in the two-dimensional system. It was found that most of the large oligonucleotides remained at the origin of the DEAE-paper, although they were separated well in the cellulose acetate ionophoresis dimension, especially when urea was added to the buffer. In order to prevent the larger oligonucleotides from being absorbed too strongly on the DEAE-paper, a high concentration of anion was necessary. This precluded the use of electrophoresis because the high ionic strength increased the conductivity, thus leading to undesirable overheating. The best results were obtained with chromatography on DEAE-paper rather than ionophoresis. A technique was developed by B. G. Barrell and R. E. Offord (unpublished work) to utilize NaCl gradient elution to displace the oligonucleotides from the DEAE-paper. Although the spots were somewhat streaky, the fractionation was usually adequate. Much of the initial work on partial digests was accomplished with this technique.

An alternative procedure called homochromatography (Brownlee *et al.*, 1968) was also developed to fractionate the larger oligonucleotide fragments. It employs a high concentration of mixed nonradioactive oligonucleotides to displace the radioactive oligonucleotides successively from the DEAE-paper. The nonradioactive nucleotides, usually obtained from a partial alkaline hydrolysis of yeast RNA, saturate the DEAE groups during the development of the chromatogram. As the solution containing the high concentration of nucleotides moves through the DEAE-paper, the nucleotides displace one another and produce a series of fronts. The smaller oligonucleotides with less affinity are displaced by the large ones and therefore move farther. The radioactive oligonucleotides move with the different fronts and are fractionated according to their affinity for the DEAE-paper. When this method of chromatography was combined with cellulose acetate electrophoresis in a two-dimensional system, it was possible to achieve separations of oligonucleotides up to 25 residues in length. Longer ones still remained at the origin of the paper (see Chapter 11).

With the aid of these procedures and techniques, separate subunits of *E. coli* 5S RNA were analyzed to yield the following five sequences:

1. pUGCCUGGCGGCCGUAG
2. GUAGCGCGGUGGUCCCACCUGACCCCAUGCC-GAACUCAGAAGUGAAACGCGUAG
3. GUAGCGCCGAUGGUAG
4. GUAGUGUGGGGUCUCCCCAUGCGAGAGUAG
5. GUAGGGAACUGCCAGGCAU-OH

All of the nucleotides present in a complete T_1 or pancreatic RNase hydrolysis of *E. coli* MRE600 5S RNA are included in these five sequences. This result represented a stage at which progress in obtaining the necessary overlapping sequences became particularly difficult. With the exception of sequence 1, all of the sequences have G-U-A-G at the left end, and all but sequence 5 have it at the right end. The problem was to determine which of these overlapped. This was eventually accomplished by a more detailed study of the larger oligonucleotides as well as those present in smaller amounts from partial RNase hydrolysis.

In particular, one oligonucleotide (T6) gave U-G, U-A-G, and U-C-U-C-C-C-A-U-G- when hydrolyzed with T_1 RNase. It also yielded G-G-G-G-U with pancreatic RNAse. This indicated that the left end of sequence 4 was involved. Other products included A-U-G from the T_1 digest, and G-G-U from the pancreatic RNase digest, thus indicating that the right end of sequence 3 was also involved. This showed that the right end of sequence 3 must overlap with the left end of sequence 4.

Another oligonucleotide (T19) was obtained from a *Bacillus subtilis* RNase hydrolysis as well as from a partial T_1 digestion. Further hydrolysis with T_1 RNase revealed the presence of A-A-A-C-G, U-A-G-, C-G, and 2 moles of C-C-G in T19. This indicated that the right-hand terminus of sequence 2 was present. Oligonucleotide T19 also gave A-G-C when digested with pancreatic ribonuclease. There are only two A-G-C sequences in this 5S RNA. One of them is near the left end of this same sequence 2 and therefore cannot be involved at the right end. The other A-G-C is at the left end of sequence 3 and thus provided the required overlap with an adjacent sequence. As a result, sequence 2 may be placed to the left of sequence 3.

The only remaining sequences are 1 and 5, representing the left and right ends, respectively. Sequences 2, 3, and 4 are therefore situated between 1 and 5 to yield: pUGCCUGGCGGCCGUAGCGCGGU-GGUCCCACCUGACCCCAUGCCGAACUCAGAAGUGAAACG-CCGUAGCGCCGAUGGUAGUGUGGGGUCUCCCCAUGCGA-GAGUAGGGAACUGCCAGGCAU$_{OH}$.

The data presented by Brownlee *et al.* (1968) and by Forget and Weissman (1967) are more than sufficient for the complete reconstruction of the correct sequence of each 5S RNA. This comparative profusion of data results as much from the use of the two-dimensional electrophoresis procedure as from the small size of the RNA: the number of different oligonucleotides found in an enzymatic digest of 5S RNA ranges between 20 and 30. Because of this, longer-chain oligonucleotides of closely similar base compositions occur infrequently, and identifications are thereby made easier.

It would appear that the prospects for the use of the two-dimensional electrophoresis procedure extend certainly through nucleic acids containing 120 nucleosides and probably beyond to molecules containing as many as 200 nucleosides (see Chapter 8). The limiting factors are the increase in the number of different oligonucleotides and in the complexity of the "fingerprints." It is the experience in most laboratories concerned with nucleic acid sequence analysis that the increase in difficulty is not arithmetic, but more nearly approaches the geometric.

Given a procedure that may be used in the sequence analysis of nucleic acid containing as many as 200 nucleotides, the problem with RNA of high molecular weight is how to cleave it into specific subunits of a suitable size. This topic will be discussed in Chapter 8.

REFERENCES

Brownlee, G. G., Sanger, F., and Barrell, B. G. (1967). *Nature*, **215**, 735.
Brownlee, G. G., Sanger, F., and Barrell, B. G. (1968). *J. Mol. Biol.*, **34**, 379.
Forget, B. G., and Weissman, S. M. (1967). *Science*, **158**, 1695.
Rosset, R., and Monier, R. (1963). *Biochim. Biophys. Acta*, **68**, 653.
Rosset, R., Monier, R., and Julien, J. (1964). *Bull. Soc. Chim. Biol.*, **46**, 87.
Sanger, F., Brownlee, G. G., and Barrell, B. G. (1965). *J. Mol. Biol.*, **13**, 373.

5.

STEPWISE ENZYMATIC PROCEDURES

One of the more direct methods proposed for the determination of base sequences in nucleic acid is stepwise enzymatic degradation. This method depends on the use of exonucleases such as spleen phosphodiesterase and venom phosphodiesterase or of polynucleotide phosphorylase to liberate nucleotides one by one from nucleic acid chains. An appropriate identification of each nucleotide is made as it is released. This is illustrated in the following schemes with the hypothetical oligonucleotide ApCpGpU:

1. Spleen phosphodiesterase (SPDE)

$$\text{ApCpGpU} \rightarrow \text{Ap} + \text{CpGpU} \rightarrow \text{Ap} + \text{Cp} + \text{GpU} \rightarrow \text{Ap} + \text{Cp} + \text{Gp} + \text{U}$$

2. Venom phosphodiesterase (VPDE)

$$\text{ApCpGpU} \rightarrow \text{ApCpG} + \text{pU} \rightarrow \text{ApC} + \text{pG} + \text{pU} \rightarrow \text{A} + \text{pC} + \text{pG} + \text{pU}.$$

3. Polynucleotide phosphorylase

$$\text{ApCpGpU} \xrightarrow{P_i} \text{ApCpG} + \text{ppU} \xrightarrow{P_i} \text{ApC} + \text{ppG} + \text{ppU} \xrightarrow{P_i} \text{A} + \text{ppC} + \text{ppG} + \text{ppU}$$

In scheme 1, SPDE catalyzes the hydrolysis of the oligonucleotide, beginning at the left end and proceeding toward the right end. The mononucleotides produced are the 3′-phosphates. With VPDE, scheme 2, the hydrolysis starts at the right end and proceeds toward the left, producing the mononucleoside 5′-phosphates. Polynucleotide phosphorylase, scheme 3, catalyzes the phosphorolysis of the oligonucleotide, starting at the right end and proceeding toward the left. The products of this reaction are the mononucleoside 5′-diphosphates.

Attempts to apply the stepwise enzymatic degradation to base sequence analysis of high molecular weight RNA have achieved only limited success. Hydrolysis of TMV RNA with VPDE (Singer and Fraenkel-Conrat, 1963) confirmed the earlier finding that adenosine occupied the right-hand terminal position in TMV RNA (Sugiyama and Fraenkel-Conrat, 1961; Whitfeld, 1962) but did not result in a reliable identification of the four or five bases adjacent to this end. Considerably more success has been obtained with VPDE in the analysis of terminal sequences to tRNA (Preiss *et al.*, 1961; Nihei and Cantoni, 1963; Zubay and Takanami, 1964) and of small oligonucleotide fragments (Holley *et al.*, 1964). The use of SPDE in the analysis of oligonucleotides has also been described (Sanger *et al.*, 1965). Although denatured DNA is susceptible to the exonucleolytic action of enzymes such as SPDE, VPDE, *E. coli* DNase I, and *B. subtilis* and *L. acidophilus* nucleases, base sequence analysis has not progressed far with this material because of the current difficulty in obtaining molecularly homogeneous samples of poly- and oligodeoxyribose nucleotides. However, it is presumed that the considerations discussed in the following section with respect to RNA may apply as well to DNA.

KINETICS OF ENZYMATIC STEPWISE DEGRADATION

Part of the problem in the use of stepwise enzymatic degradation in base sequence analysis of high molecular weight nucleic acid appears to be the selection of optimum conditions for the reaction. A theoretical study has been made by Cantor *et al.* (1964) of exonuclease kinetics and its application to the determination of base sequences in nucleic acids. It was concluded that steady-state experimental conditions provide the poorest resolution in the liberation of successive nucleotides, and that the use of high enzyme concentrations, of the order of the Michaelis constant, is desirable.

These conclusions resulted from the following basic assumptions about the nature of exonucleases (including polynucleotide phosphorylase) and their substrates:

1. The enzymes of interest have no base specificity: the overall rate of liberation of monomers is independent of sequence or base composition.
2. The rate of degradation of oligomers is independent of chain length.
3. Either the substrates are single-strand polymers or the exonucleases are not dependent on conformation.
4. The population of nucleic acid polymers is molecularly homogeneous; there are no spurious ends.
5. The enzyme is not contaminated with other nuclease or nucleic acid-modifying activity.

The irreversible degradation of a polynucleotide chain, from one end exclusively, is considered. A representation of the sequence of steps is shown for a random attack by the enzyme on the substrate molecules:

$$\mathrm{E} + \mathrm{S_N} \underset{k_{-1}}{\overset{k_1}{\rightleftharpoons}} \mathrm{ES_N} \xrightarrow{k_2} \mathrm{E} + \mathrm{S_{N-1}} + \mathrm{B_1}$$

$$\vdots$$

$$\mathrm{E} + \mathrm{S}_{\mathrm{N}-n+1} \underset{k_{-1}}{\overset{k_1}{\rightleftharpoons}} \mathrm{ES}_{\mathrm{N}-n+1} \xrightarrow{k_2} \mathrm{E} + \mathrm{S}_{\mathrm{N}-n} + \mathrm{B}_n$$

$$\vdots$$

$$\mathrm{E} + \mathrm{S_2} \underset{k_{-1}}{\overset{k_1}{\rightleftharpoons}} \mathrm{ES_2} \xrightarrow{k_2} \mathrm{E} + \mathrm{S_1} + \mathrm{B_{N-1}}$$

$\mathrm{S_N}$ is the undegraded nucleic acid, $\mathrm{B_1}$ is the first nucleotide liberated, and only a single enzyme-substrate complex, $\mathrm{ES}_{\mathrm{N}-n}$, is present for each degradation step.

The kinetics for the scheme of stepwise degradation may be represented by a set of differential equations for (a) the polymeric chains,

$$(\dot{S}_N) = \frac{d(S_N)}{dt} = -k_1(E)(S_N) + k_{-1}(ES_N)$$

$$(\dot{S}_{N-n}) = -k_1(E)(S_{N-n}) + k_{-1}(ES_{N-n}) + k_2(ES_{N-n+1}) \qquad (1)$$

$$(\dot{S}_1) = k_2(ES_2)$$

(b) the enzyme-substrate complexes,

$$(\dot{E}S_{N-n}) = k_1(E)(S_{N-n}) - (k_{-1} + k_2)(ES_{N-n}) \qquad (2)$$

and (c) the liberated nucleotides,

$$(\dot{B}_n) = k_2(ES_{N-n+1}) \qquad (3)$$

The final step produces two nucleotides, B_{N-1} and $S_1 = B_N$.

This set of nonlinear equations can be linearized and solved by assuming that the concentration of free enzyme (E) is substantially independent of time. Such would be the case when the enzyme concentration greatly exceeds the concentration of the reactive polynucleotide ends. The results of these solutions are expressions for the concentrations of the various polymeric chains and liberated nucleotides. These expressions are, in general, polynomial sums in powers of t multiplied by two exponentials $e^{\alpha_1 t}$ and $e^{\alpha_2 t}$, where α_1 and α_2 are the two roots of the quadratic equation

$$\alpha_{1,2} = \frac{-(k + k_{-1} + k_2) \pm ((k + k_{-1} + k_1)^2 - 4kk_2)^{1/2}}{2} \qquad (4)$$

in which $k = k_1(E)$. The presence of only two distinct exponential factors is a consequence of the assumption of a single enzyme-substrate complex for each step. Additional complexes would introduce more rate constants and exponential factors.

Where there is a steady-state concentration of the various enzyme-substrate complexes, $k \ll k_{-1}$ or k_2, or $k \gg k_{-1}$ or k_2, expansion of the square root in equation (4) with retention of the first two terms yields

$$\alpha_1 \simeq -\frac{kk_1}{k + k_{-1} + k_2} \qquad (5)$$

$$\alpha_2 \simeq -(k + k_{-1} + k_2) \qquad (6)$$

Equation (5) may be written in the more familiar terms of steady-state enzyme kinetics as

$$\alpha_1 \simeq -\frac{V_m/(\mathrm{E})_0}{1 + K_m/(\mathrm{E})} \tag{7}$$

where the maximum velocity $V_m = k_2(\mathrm{E})_0$, the Michaelis constant $K_m = (k_{-1} + k_2)/k_1$, and $(E)_0$ is the total enzyme concentration. Considering that in most conventional kinetic studies $K_m \gg (E)_0 > E$, Cantor derived the relationship

$$\alpha_1 \simeq -\frac{V_m(\mathrm{E})}{K_m(\mathrm{E})_0} = -\frac{V_m}{K_m}\left[\frac{1}{1 + \mathrm{S_N}(0)/K_m}\right] \simeq -\beta \tag{8}$$

where $\mathrm{S_N}(0)$ is the initial concentration of substrate. Because $|\alpha_2| \gg |\alpha_1|$, the transient terms involving $e^{\alpha_2 t}$ die out much more rapidly than the steady-state terms involving $e^{\alpha_1 t}$. Therefore the steady-state solution for the concentration of a given nucleotide is

$$\mathrm{B}_n = \mathrm{S_N}(0)\left[1 - \sum_{i=0}^{n=1} \frac{(\beta t)^i e^{-\beta t}}{i!}\right] \tag{9}$$

In the non-steady-state situation, two cases arise when (a) k and $k_2 \gg k_{-1}$, and (b) $k \simeq k_1(\mathrm{E}) \simeq k_2$. From equation (4),

$$\alpha_1 \simeq \alpha_2 \simeq -k_2 \simeq -\alpha$$

and the concentration of the nucleotide is given by

$$\mathrm{B}_n(t) = \mathrm{S_N}(0)\left[1 - \sum_{i=0}^{2n-1} \frac{(\alpha t)^i e^{-\alpha t}}{i!}\right] \tag{10}$$

Each of the cases in the non-steady-state situation requires a total enzyme concentration of at least the same order of magnitude as K_m because

$$\frac{k_2}{k_1\mathrm{E}} \geqq \frac{K_m}{(E)_0}$$

The question whether the steady-state or the non-steady-state condition should be used in the stepwise sequence analysis of nucleic acids is resolved by calculations of the time of appearance of nucleotides under each of the two sets of conditions. The results for the 1st, 3rd, 5th, and 25th bases are plotted in Fig. 5.1. It can be seen that the nucleotides

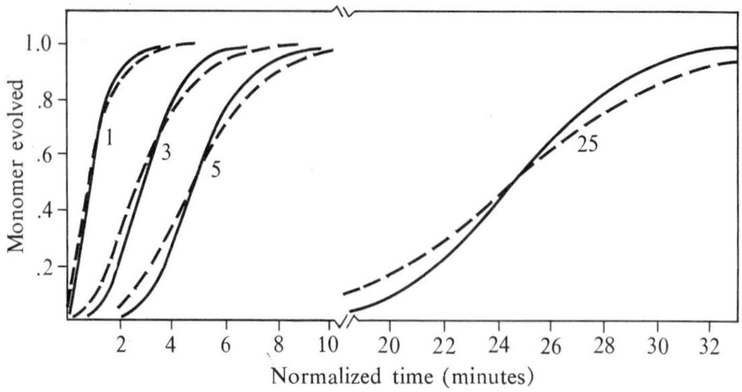

Fig. 5.1. Time of appearance of mononucleotides under two different sets of kinetic conditions. ——— = non-steady-state situation where $\alpha_1 \cong \alpha_2$ and total enzyme concentration $\cong K_m$. - - - - = steady-state situation where the total enzyme concentration $\ll K_m$.

are liberated in a more sharply defined time interval when $\alpha_1 = \alpha_2$ than for the steady-state condition. This difference can also be expressed in the form of a dimensionless sharpness parameter, P_n,

$$P_n = \frac{\text{time when } \mathrm{B}_n = 0.9 - \text{time when } \mathrm{B}_n = 0.1}{\text{time when } \mathrm{B}_n = 0.5 - \text{time when } \mathrm{B}_{n-1} = 0.5}$$

The smaller the P_n, the greater is the resolution. Various values of P_n have been calculated for a number of kinetic parameters. They are shown in Table 5.1. When $\alpha_1 = \alpha_2$, the values for P_n are lower than when the steady-state parameter is used. Figure 5.1 is a plot of the values obtained when P_n is calculated for $\alpha_1 = \alpha_2$ and the steady-state condition. As before, the case when $\alpha_1 = \alpha_2$ provides the clearest resolution.

APPLICATIONS OF KINETICS TO SEQUENCE ANALYSIS

Because of the lack of appropriate exonuclease data, Cantor *et al.* (1964) elected to test the kinetic approach through use of a hypothetical 12-mer,

C-U-U-G-A-G-G-U-A-C-C-A

Table 5.1. Effect of kinetic parameters on P_n

Rate Constants				P_n				
k	k_{-1}	k_2	Monomer (n) =	1	2	3	4	5
Steady state				2.2	3.4	4.2	4.9	5.6
10^{-3}	1	1		2.2	3.4	4.2	5.0	5.6
10^{-4}	1	1		2.2	3.4	4.2	5.0	5.6
10^{3}	1	1		2.2	3.4	4.2	5.0	5.6
10^{3}	1	10^{-4}		2.2	3.4	4.2	5.0	5.6
1	1	1		2.2	3.3	3.9	4.5	4.9
10^{3}	10^{-3}	10^{2}		2.2	3.3	4.0	4.6	5.2
1.5	0.16	1.3		1.8	2.7	3.3	3.7	4.3
1100	10^{-3}	10^{3}		1.8	2.6	3.1	—	—
$\alpha_1 = \alpha_2$				1.7	2.5	3.1	3.6	4.0

Assuming the kinetic parameters of near optimal conditions, $k = 1.5$, $k_2 = 1.3$, $k_{-1} = 0.16$, and $\alpha_1 = \alpha_2 = 1$, calculations were made and curves were constructed which related the concentration of each nucleotide evolved to the time of exonuclease (SPDE) action. This is illustrated in Fig. 5.2. The task, then, is to infer the correct sequence from these data through use of kinetics.

An initial inspection of the curves shows that each of the four bases reaches a maximum concentration of 3, thus establishing the base composition of the hypothetical oligonucleotide as $A_3C_3G_3U_3$. Next, calculation of the curves corresponding to the various B_n from equation (10) reveals that B_1, B_2, B_4, and B_5 coincide with C, U, G, and A, respectively, up to a concentration of 0.8. The indicated sequence is therefore

C-U- -G-A- - - - - - -

The last base to approach its maximum concentration is A, indicating that it occupies the right-hand terminus:

C-U- -G-A- - - - - - -A

Further inspection of the curve suggests that the third base is U because its concentration is almost 1.5 before any appreciable amount of the fourth base, G, appears. The sequence may therefore be written

C-U-U-G-A- - - - - - -A

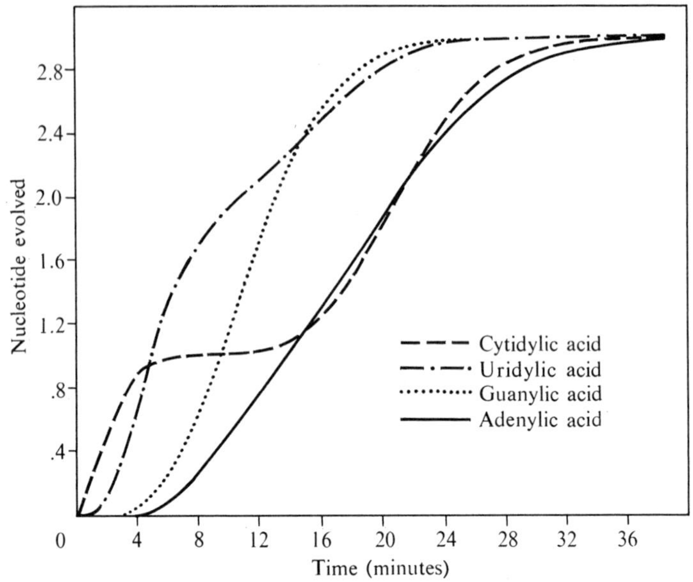

Fig. 5.2. Calculated curves of the time of appearance of mononucleotide from exonuclease action under non-steady-state conditions. Hypothetical 12-mer = C–U–U–G–A–G–G–U–A–C–C–A.

An inflection point appears in the curve for the evolution of U at $t = 15$. From equation (10), the second derivative, $\ddot{B}_n = 0$, when $t = (2n - 1)/\alpha$. Since $\alpha = 1$, position 8 must be occupied by U:

C-U-U-G-A- - -U- - - -A

thus accounting for all of the U. Figure 5.2 also shows that all of the G's appear before all of the U's are liberated. This indicates that the remaining two G's must occur in positions before the final U, which is in position 8. The only positions available are 6 and 7:

C-U-U-G-A-G-G-U- - - -A

This leaves two C's and one A unaccounted for. There are three possible sequences for the last four bases at the right-hand terminus: (a) -C-C-A-A-, (b) -A-C-C-A, and (c) -C-A-C-A. In Fig. 5.2, at time $t = 16$–20, the concentration of C is seen to fall below that of A. This

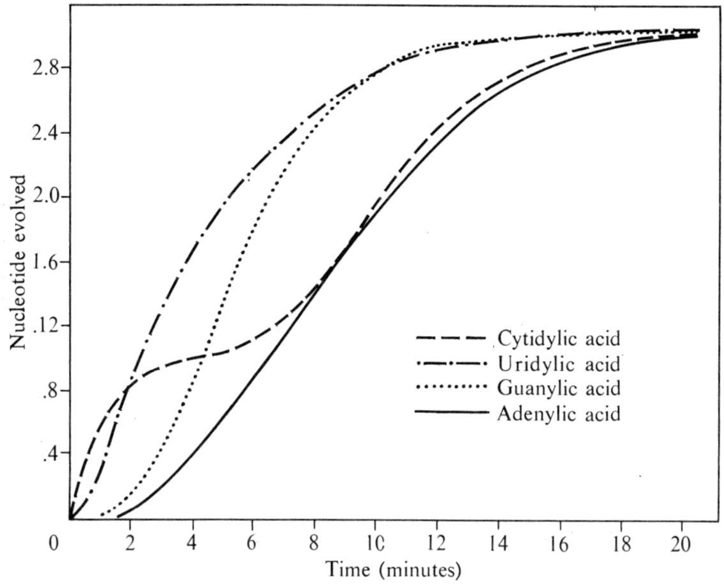

Fig. 5.3. Calculated curves of the time of appearance of mononucleotides from exonuclease action under steady-state conditions. Hypothetical 12-mer as in Fig. 5.2.

eliminates possibilities (a) and (c) because they would require the concentration of C to exceed A always. The inferred sequence is therefore

C-U-U-G-A-G-G-U-A-C-C-A

It would be much more difficult to reconstruct a correct sequence from data obtained under steady-state conditions. Calculations of the change of base concentration with time made with equation (9) are plotted in Fig. 5.3. It can be seen that the inflection in the curve for U, the intersection of the A and C curves, and the details of the four curves near the asymptotes are virtually obscured. Even under optimum kinetic conditions, it may be impossible, owing to experimental error, to make distinctions such as those between C and A in the late stage of the reaction plotted in Fig. 5.2. It has been estimated that 1% is the maximum amount of total error that can be tolerated in the stepwise degradation of a 20-nucleotide sequence.

Pending substantial improvements in experimental manipulations and measurements, it would appear that the application of stepwise enzymatic degradation is limited to the determination of sequences in the smaller oligonucleotides.

PARTIAL HYDROLYSIS WITH EXONUCLEASE

As mentioned before, SPDE and VPDE have been used successfully, under conditions where $K_m \gg (E)_0$, to determine the sequence of short oligonucleotides. The rationale for this approach depends on the fact that oligonucleotides of any one chain length are identical when produced by exonucleolytic action. In the following scheme, this is illustrated by the VPDE partial hydrolysis of a hypothetical oligonucleotide ApCpGpU.

$$\text{ApCpGpU} \xrightarrow{\text{VPDE}} \begin{array}{l} \text{ApCpG} + \text{pU} \\ \text{ApC} + \text{pG} + \text{pU} \\ \text{A} + \text{pC} + \text{pG} + \text{pU} \end{array}$$

Fractionation and analysis of the partial hydrolytic products in the mixture would show that the nucleoside portion consisted entirely of adenosine. Because of the specificity of VPDE, this establishes A as the left end of the oligonucleotides. The mononucleotide fraction would consist of a mixture of all of the mononucleotides released by the exonucleolytic action. The dinucleotide, trinucleotide, and tetranucleotide fractions would consist only of ApC, ApCpG, and ApCpGpU, respectively. The original sequence could be reconstructed from the data obtained by further hydrolysis of the isolated oligonucleotides with alkali or SPDE. In each case, complete hydrolysis should result in a mixture of mononucleotides and a single nucleoside. This nucleoside would represent the right-hand terminus of each oligonucleotide:

$$\text{ApCpGpU} \xrightarrow[\text{SPDE}]{\text{KOH or}} \text{Ap} + \text{Cp} + \text{Gp} + \text{U} \quad \text{(fourth nucleoside)}$$

$$\text{ApCpG} \xrightarrow[\text{SPDE}]{\text{KOH or}} \text{Ap} + \text{Cp} + \text{G} \quad \text{(third nucleoside)}$$

$$\text{ApC} \xrightarrow[\text{SPDE}]{\text{KOH or}} \text{Ap} + \text{C} \quad \text{(second nucleoside)}$$

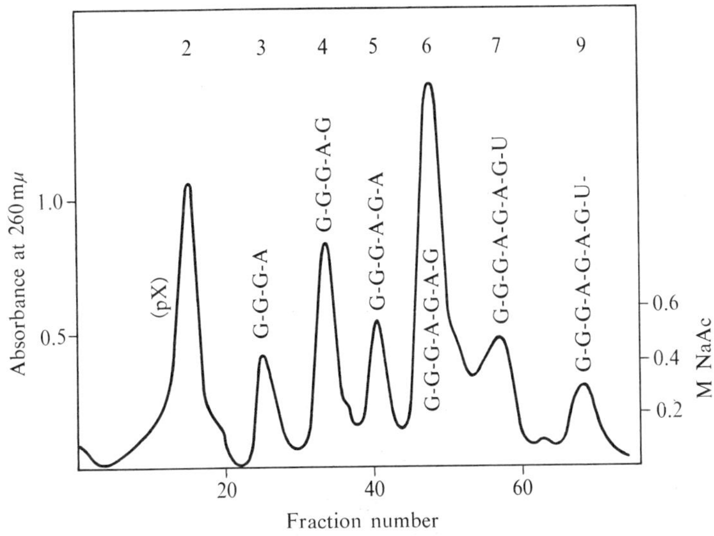

Fig. 5.4. Chromatographic separation of products from a partial SPDE digestion of GpGpGpApGpApGpUp. Chromatography performed with DEAE-cellulose column-7 *M* urea system and a linear salt gradient. Redrawn from Holley *et al.* (1964).

The sequence of the original oligonucleotide to the right of the left terminus, Ap, is established by adding the nucleoside produced from each successively smaller oligonucleotide to the right terminal nucleoside U:

$$A + C + G + U = ApCpGpU$$

In this manner, Holley and co-workers (1964) were able to establish the sequence of the octanucleotide GpGpGpApGpApGpUp (see Table 3.4). Partial hydrolysis with SPDE and separation of the reaction mixture on a DEAE-cellulose column in 7 *M* urea yielded the pattern shown in Fig. 5.4. The terminal phosphate was first removed from the oligonucleotide by treatment with phosphomonoesterase. Peak 9 in the chromatogram represented a small amount of the original phosphorylated octanucleotide, whereas peak 7, with two less negative charges, contained the dephosphorylated octanucleotide. Peaks 6, 5, 4, and 3 were successive degradation products. Upon alkaline hydrolysis, the material in each peak yielded only one nucleoside. They were guanosine, adenosine, guanosine, and adenosine, respectively, for peaks 6, 5, 4, and 3. Because the octanucleotide was known to contain two Ap's, five Gp's, and one Up, with Up at the right end and one of the Gp's at the left end, the sequence could be reconstructed as

G-G-G-A-G-A-G-U

With SPDE, partial hydrolysis also yields successively smaller oligonucleotides, as shown in the reaction with a hypothetical oligonucleotide ApCpGpU:

$$\text{ApCpGpU} \xrightarrow{\text{SPDE}} \text{Ap} + \text{CpGpU}$$
$$\text{Ap} + \text{Cp} + \text{GpU}$$
$$\text{Ap} + \text{Cp} + \text{Gp} + \text{U}$$

A fractionation and analysis of the mixture would show that the nucleoside portion consisted entirely of uridine, thus establishing U as the right-hand end of the oligonucleotide. Hydrolysis of each successively smaller oligonucleotide with VPDE would release a nucleoside from which the original sequence could be reconstructed:

$$\text{ApCpGpU} \xrightarrow{\text{VPDE}} \text{A (first nucleoside)} + \text{pC} + \text{pG} + \text{pU}$$
$$\text{CpGpU} \xrightarrow{\text{VPDE}} \text{C (second nucleoside)} + \text{pG} + \text{pU}$$
$$\text{GpU} \xrightarrow{\text{VPDE}} \text{G (third nucleoside)} + \text{pU}$$

In practice, the nature of the mononucleotide released to form the successively smaller oligonucleotides may be determined by a novel electrophoretic method developed by Sanger and co-workers (1965). The procedure involves the partial digestion of an unknown oligonucleotide with SPDE, followed by separation of the products by electrophoresis on DEAE-paper at pH 1.9 (see Chapter 11). It was found that the distances traveled by the successively smaller oligonucleotides were related to their sequences. For example, consider the tetranucleotide ApUpCpGp: partial digestion with SPDE should produce

UpCpGp + CpGp + unhydrolyzed ApUpCpGp
+ Ap + Up + Gp + Gp

Table 5.2. *M* values for nucleotides from T_1 RNase digests

Left Terminus	M
C	0.05–0.3
A	0.5 –1.0
U	1.5 –2.5

Upon electrophoresis on DEAE-paper at pH 1.9, the mononucleotides travel farthest, followed by the dinucleotide, the trinucleotide, and the tetranucleotide, in that order. If we let y equal the distance traveled from the origin by the unhydrolyzed ApUpCpGp, and x equal the distance between ApUpCpGp and its first degradation product UpCpGp, a value $M = x/y$ can be defined for ApUpCpGp that depends on the nature of the left terminus A. It was found that, for oligonucleotides from T_1 RNase digests, the values of M lie within the limits shown in Table 5.2 with no overlapping between the values for the different residues. Similarly, the distance traveled from the origin by the first degradation product UpCpGp could be set equal to y, and the distance between it and the second degradation product CpGp could be set equal to x. The value M could then be used to identify the left terminal nucleotide, Up, of the first degradation product.

As an example of the application of this method, a partial hydrolysis of a particular tetramer followed by electrophoretic separation might produce the pattern of spots shown in Fig. 5.5. In this example, the distance of the unhydrolyzed tetramer sample from the origin is 1.1 cm. The distance between the unhydrolyzed tetramer and the first degradation product is 0.8 cm, so that $M = 0.8/1.1 = 0.7$. From Table 5.2

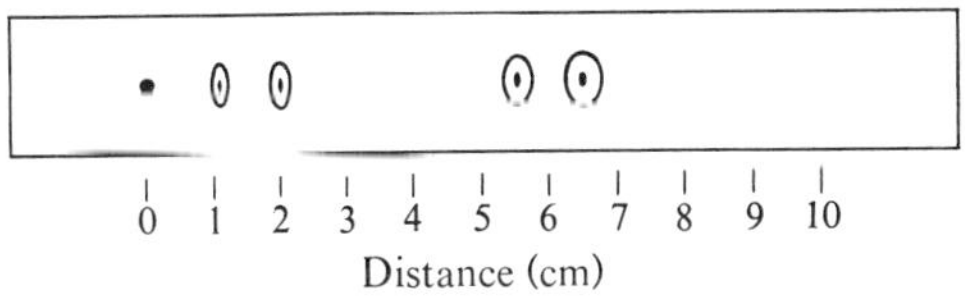

Fig. 5.5. Electrophoretic separation of products from a partial SPDE hydrolysis of hypothetical tetramer ApUpCpGp. All of these oligonucleotides would be expected to migrate more slowly than the blue marker dye when electrophoresis was performed on DEAE-paper in 7% formic acid.

an *M* value of 0.7 is in the range for Ap; hence the left terminal nucleotide of the unhydrolyzed tetramer is Ap. The next nucleotide is identified in the same way. The distance of the first degradation product from the origin is 1.9 cm. The distance between the first and second degradation products is 3.6 cm. This yields an *M* value of 1.9, thus identifying the second nucleotide from the left end as Up. Completion of the sequence determination is accomplished by measurement of x and y for the third degradation product or its identification as CpGp from the known chromatographic behavior, or by a base compositional analysis of the original tetranucleotide. The data establishes the sequence as ApUpCpGp.

The factor that tends to limit the effectiveness of this method most is the varying rates at which the different nucleotides are split off by SPDE. Under the conditions used by Sanger and co-workers, Up is split off more rapidly than Ap, which in turn is split off more rapidly than Cp. Therefore some of the possible degradation products may not appear in a particular digest, and this could lead to a misinterpretation of the results, especially in the cases of larger oligonucleotides.

With the shorter oligonucleotides from pancreatic RNase digests, the problem is not a difference in rates of hydrolysis, since Ap and Gp are split off at similar rates. A problem does arise from the fact that the *M* values for Ap range from 0.4 to 1.1, while those for Gp range from 1.2 to 3.1. Because the spread of the values is fairly wide and the limits are close, it is advisable to seek confirmation from other data when *M* falls in the range 0.9 to 1.5. Fortunately, the identity of the shorter oligonucleotides may often be confirmed by a comparison with the electrophoretic position of authentic samples.

In common with most other sequence analysis procedures, this method is sensitive to the introduction of extraneous breaks in the oligonucleotide chain. As always, care should be taken to minimize the effects of unwanted endonucleolytic and thermal degradation.

REFERENCES

Cantor, C. R., Tinoco, I., Jr., and Peller, L. (1964). *Biopolymers*, **2**, 51.

Holley, R. W., Madison, J. T., and Zamir, A. (1964). *Biochem. Biophys. Res. Commun.*, **17**, 389.

Nihei, T., and Cantoni, G. L. (1963). *J. Biol. Chem.*, **238**, 3991.

Preiss, J., Duckmann, M., and Berg, P. (1961). *J. Biol. Chem.*, **36**, 1748.
Sanger, F., Brownlee, G. G., and Barrell, B. G. (1965). *J. Mol. Biol.*, **13**, 373.
Singer, B., and Fraenkel-Conrat, H. (1963). *Biochim. Biophys. Acta*, **72**, 534.
Sugiyama, T., and Fraenkel-Conrat, H. (1961). *Proc. Natl. Acad. Sci.*, **47**, 1393.
Whitfeld, P. R. (1962). *J. Biol. Chem.*, **237**, 2865.
Zubay, G., and Takanami, M. (1964). *Biochem. Biophys. Res. Commun.*, **15**, 207.

6.

STEPWISE CHEMICAL DEGRADATION

A cardinal consideration in any sequence analysis by stepwise degradation is the synchrony of each step. In Chapter 5 it was shown that, for stepwise enzymatic degradation under optimum kinetic conditions, very careful control must be exercised in experimental manipulations in order to determine more than a few sequences in RNA of high molecular weight. Real kinetic parameters appear to limit the ability to establish and maintain a sharp synchrony for the release of each base in enormous populations of reacting enzymes and substrates.

This problem would be virtually solved if at any one step it were possible to prevent further degradation of nucleic acid until all of a particular base had been released. Just such a scheme was suggested by Brown *et al.* (1953) and by Whitfeld (1954), based on an observation of Whitfeld and Markham (1953). The principal features of the scheme as described by Whitfeld (1954) remain essentially the same today and are illustrated in Fig. 6.1. Structure (a) depicts the last three nucleotides at the right-hand end of a polynucleotide. In the first step, the removal of the terminal phosphate is catalyzed by the enzyme phosphomonoesterase (PME) to yield the structure (b) and inorganic phosphate. When this reaction is essentially complete, the polynucleotide is treated with a dilute solution of periodate at pH 5. Under these conditions, the 2′ and 3′ (vicinal) hydroxyls of the terminal nucleoside are rapidly and specifically oxidized to a dialdehyde, structure (c). In the next step the

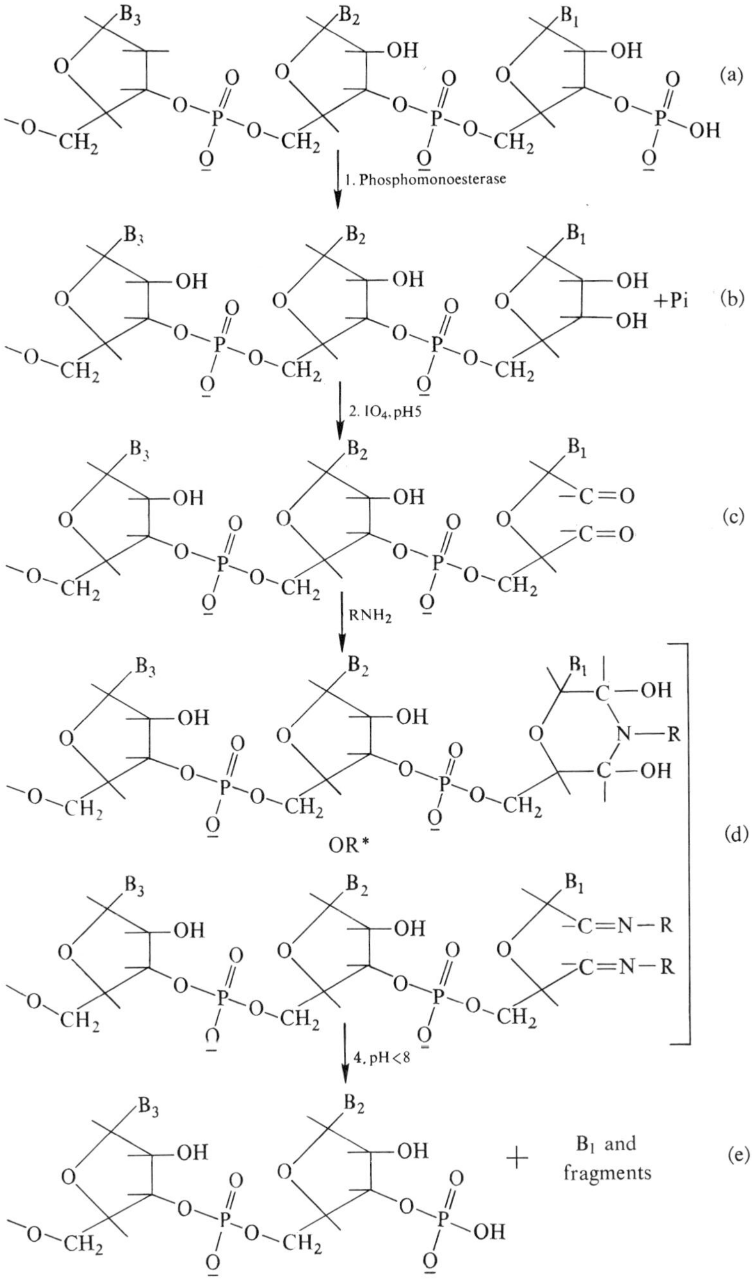

Fig. 6.1. Chemical stepwise degradation of RNA. (* True structure of intermediates not established.)

dialdehyde reacts with an amine to form an addition product that probably has the structure shown in the upper portion of (d) but has been represented as the lower structure. At pH below 8 and in the presence of excess amine, the addition product undergoes a β-elimination to yield the terminal base, fragments of the dialdehyde, and the remainder of the polynucleotide with a 3′-phosphomonoester at the new right-hand terminus. A new cycle may be initiated by removal of the phosphomonoester with PME from the right-hand terminal nucleoside of structure (e). After this, the other steps are repeated in the same order as the first cycle.

It is plain that each of the reaction steps in any one cycle involves only the terminal nucleotide, and that the penultimate nucleotide can participate only after the terminal nucleotide has been removed, and then only after the blocking terminal phosphomonoester has been removed. In theory, then, this procedure provides an excellent means for achieving the synchronous stepwise degradation of nucleic acids. application of this scheme to the sequence analysis of more than a few bases requires that certain conditions be met. They are:

1. The nucleic acid to be analyzed must have a high degree of molecular homogeneity. Impurities contribute to a high background which makes identification of the base released at the end of each cycle less certain.

2. Generation of spurious ends must be minimized during the analysis. The seriousness of this problem may be appreciated if one considers the occurrence of a single extraneous break per molecule per cycle in a chain of 1000 nucleotides. By the fourth cycle, the true end is experimentally not distinguishable from the false ends.

Sources of most of the spurious ends, or noise, are thermal degradation and extraneous nuclease action. Little can be done about the thermal degradation except to carry out the reaction at reduced temperatures, if necessary.

The presence of small amounts of extraneous nuclease is probably responsible for the generation of most of the spurious ends. Precautions against the introduction of nuclease during experimental manipulations are essential, and the use of a highly purified, nuclease-free PME is absolutely necessary. Preparations of nuclease-free PME may be obtained from nuclease-deficient mutant *E. coli* A19 or a commercial preparation (Worthington) of chromatographically purified *E. coli* alkaline phosphatase. All such preparations should be checked for nuclease activity by the release of inorganic phosphate and by the change in apparent molecular weight when the PME is incubated with RNA of high molecular weight.

Other sources of contaminating nucleases include bacteria and the "finger nucleases" (Holley *et al.*, 1961). Ordinarily, all glassware should be treated with chromic acid cleaning solution and heated at 170° in an oven for 6 hours. All solutions should be sterilized if possible, and the use of sterile, disposable plastic gloves for laboratory personnel is recommended.

3. The amount of starting material must be sufficient to permit the cyclic analysis to continue for the desired chain length. In general, the amount needed is expressed by

$$A = \frac{S}{F^n}$$

where F is the fraction of polymer recovered after each step, n is the chain length, and S is the minimum amount of base that can be identified after each cycle. It is evident that A can be an impossibly large number for nucleic acids of more than 1000 bases, even when the fraction of nucleic acid recovered after each cycle is 0.99. This brings us to the consideration of the yield from each step.

4. The yield from each step should be as nearly complete as possible, not only from the standpoint of the amount of material needed as indicated above, but also because of the possibility that the accumulated partially reacted ends will contribute incorrect bases in subsequent cycles of degradation. In most of the recent studies with this method, the reported yields for the removal of the terminal phosphate and the oxidation with periodate have been essentially complete.

The results with the amine-catalyzed β-elimination reaction appeared to vary according to the nature of the amine and the conditions selected for the cleavage. As an example, it was found that all of the liberated base could be recovered if excess periodate were present during the amine cleavage reaction (Neu and Heppel, 1964). Complete removal of periodate before the amine cleavage reaction resulted in the production of a mixture of the free base and the base associated with a sugar

Table 6.1. Effect of amines on cleavage reaction

	Amine	Extent of Cleavage 24°, %	45°, %
1. Values obtained for 90-minute incubation of amine with periodate-oxidized ApA (Neu and Heppel, 1964).	Lysine	32	95
	Methylamine	58	98
	Cyclohexylamine	80	98
2. Value from elimination of terminal nucleoside of periodate-oxidized TMV RNA, pH 5, 180-minute incubation, 25° (Steinschneider and Fraenkel-Conrat, 1966).	Aniline	98	--

fragment (aglycone). Alternatively, the stoichiometric amount of base could be released by treatment of this mixture with acid. A number of different amines have been suggested for use in this reaction; some of them are listed in Table 6.1. Also shown in the table is the extent of the reaction when periodate-treated oligonucleotides or nucleic acid reacts with the amines. It can be seen that the most complete reaction, under the mildest conditions, occurs when aniline is used. Although the more vigorous conditions are effective in the case of oligonucleotides, the milder conditions are preferred for the determination of base sequences in RNA of high molecular weight such as intact TMV RNA. With RNA of high molecular weight it has been found that very few internally undegraded molecules survive five or six cycles (Whitfeld, 1965) even under the mildest conditions (Steinschneider and Fraenkel-Conrat, 1966).

APPLICATION OF STEPWISE DEGRADATION

A summary of the results obtained from a stepwise base sequence analysis of TMV RNA is shown in Table 6.2. It can be seen that there is general agreement about the identity of the terminal base as adenine, and that the sharper distinctions between the bases are made in experiments 2 and 3, which were carried out under the milder conditions with aniline. Experiments 2 and 3 support the sequence (-Gp)CpCpCpA for the right end of TMV RNA, with some doubt about the fourth and fifth positions because of the less clear results from the fourth and fifth

Table 6.2. Stepwise degradation of TMV RNA

		Moles Found		
Degradation Step	Base	Exp. 1[a]	Exp. 2[b]	Exp. 3[b]
I	Adenine	0.59	0.71	0.85
	Guanine	0.11	0.04	0.05
	Uracil	0.12	0.03	0.05
	Cytosine	0.10	0.03	0.02
II	Adenine	0.42	0.14	0.12
	Guanine	0.14	0.03	0.09
	Uracil	0.11	0.07	0.07
	Cytosine	0.55	0.33	0.59
III	Adenine	0.30	0.18	0.13
	Guanine	0.11	0.08	0.06
	Uracil	0.15	0.16	0.14
	Cytosine	0.28	0.92	0.72
IV	Adenine	0.54	0.13	—
	Guanine	0.19	0.19	0.19
	Uracil	0.28	0.10	0.29
	Cytosine	0.37	0.43	0.62
V	Adenine	[c]	0.20	0.27
	Guanine	—	0.39	0.44
	Uracil		0.13	0.33
	Cytosine		0.11	0.48

[a] Values from Whitfeld (1965). Methylamine used at 30°, 4 hour incubation.
[b] Values for Exp. 2 and Exp. 3 are from Steinschneider and Fraenkel-Conrat (1966). Aniline used at 25°, 3 hour incubation.
[c] RNA was extensively degraded.

cycles. That GpCpCpCpA is in fact the sequence at the right terminus has been confirmed by an independent method (Mandeles, 1967).

Stepwise chemical degradation has also been used to determine the right-hand end sequence of bacteriophage f2 RNA (Weith and Gilham, 1967) and Qβ RNA (Weith and Gilham, 1969). In the first instance, the terminal fragment from a T_1 RNase digest of f2 RNA was first isolated through use of the same periodate oxidation step and absorption to a DEAE-cellulose column. The terminal base, adenine, was eliminated during the recovery of this fragment, leaving an oligonucleotide with the base composition $C_5U_2A_2$. Because this fragment

Table 6.3. Stepwise degradation of octanucleotide from f2 RNA

Cycle	Bases Released (mμmoles)				Cycle	Bases Released (mμmoles)			
	C	U	A	G		C	U	A	G
1	109	<2	<5	<3	4	<4	<2	6	<3
2	62	<2	<3	<3	5	17	<2	<3	<3
3	43	<2	6	<3	6	15	<2	<1	<3

was produced originally from a T_1 RNase hydrolysis, the structure deduced was

-G-(C_5-U_2-A_2)-A

A portion of the 9-mer was submitted to stepwise analysis, and the results are shown in Table 6.3. The identities of the first six bases to the left of the terminal A are quite clearly shown as

-C-C-A-C-C-C

and the terminal sequence as

-G(U_2,A)-C-C-A-C-C-C-A

A pancreatic RNase hydrolysis of another portion of the 9-mer revealed the presence of 2Up, 3Cp, and 2 ApCp. The two ApCp oligonucleotides could arise only if the terminal sequence was

-G-U-U-A-C-C-A-C-C-C-A

thus completing the sequence analysis of the right end of f2 RNA.

For the analysis of the right end fragment of Qβ RNA (Weith and Gilham, 1969), essentially the same procedure was used, except for changes in the buffer system and the chromatographic matrix. The base composition of this right end fragment (minus the terminal A as before) was $C_{10}U_4$. After 10 cycles of periodate oxidation-β-elimination, the order of the bases was established as

(C_3U)-C-U-C-U-C-C-U-C-C-C-

The sequence of the portion in the parenthesis was determined by alkaline hydrolysis and chromatography. Uridine and cytidylic acid were obtained in the ratio 3 to 1. Considering that the original fragment was obtained from a T_1 RNase digestion of Qβ RNA, the end sequence indicated is

G-C-C-C-U-C-U-C-U-C-C-U-C-C-C-A

With tRNA, as many as 19 consecutive degradations have been carried out by Uziel and Khym (1969). This sequence analysis was performed on phenylalanine tRNA from *E. coli* B (see Uziel and Gassen, 1969). One distinctive feature of this procedure was the use of the special solubility properties of a cetyltrimethylammonium (CTA) salt of the nucleic acid (Khym and Uziel, 1968). It was found that nucleic acids readily form complexes with CTA that are soluble in concentrated (1 *M*) chloride solutions, but are insoluble in more dilute solutions (0.3 *M*) (Dutta *et al.*, 1953).

This property was used to precipitate quantitatively the parent nucleic acid from the solution containing the eliminated terminal base and the PME.

In summary, the recommended procedure involves:

1. Reaction of the nucleic acid-CTA complex with a solution containing a mixture of lysine and periodate at pH 8.1. This treatment results in the oxidation of the vicinal hydroxyls, followed by the elimination of the terminal base.

2. Addition of ethylene glycol to reduce the excess periodate, followed by dilution of the mixture to precipitate the remaining nucleic acid-CTA complex. The suspension is centrifuged to remove the nucleic acid-CTA precipitate, and this precipitate is washed and saved for the next step. The supernatant and the washings are combined and analyzed for the presence (and identity) of the base eliminated in step 1.

3. The nucleic acid-CTA precipitate is dissolved in 0.2 *M* lysine and treated with PME to remove the new terminal phosphate.

4. At the conclusion of the dephosphorylation reaction, the nucleic acid-CTA complex is precipitated by dilution with water and recovered by centrifugation. The washed precipitate is then ready for solution and reaction with the lysine-periodate mixture in step 1.

The principal features of this scheme are the use of oxidation and base elimination in the same step and the simplified manipulation of the nucleic acid by virtue of the solubility properties of the CTA complex.

In carrying out this procedure, the authors advise close adherence to the conditions specified for each step.

FUTURE PROSPECTS

What are the prospects for the use of stepwise chemical degradation in base sequence determinations? It would appear to be effective in the analysis of oligonucleotides where a particular sequence or conformation makes the use of enzymatic methods difficult or unreliable. The chemical method shows no signs of the specificity that is characteristic of enzyme reactions.

In the case of high molecular weight RNA, the extensive internal degradation that is observed after five or six cycles seriously limits the method. It is probable that the introduction of extraneous nucleases and the less than optimal choice of reagents and experimental conditions contribute to the generation of spurious ends. The problem is how to eliminate as many of these causes as possible in order to increase the extent of unequivocal base sequence analysis.

AUTOMATIC SEQUENCE ANALYSIS

At present, the procedure is to oxidize the RNA by treatment with dilute periodate in the dark at 25° for 20 to 30 minutes. The reaction is terminated by the addition of ethylene glycol to exhaust the excess periodate. An amine is then added and the incubation at 25° is continued for a few hours. At the end of this period of incubation, the remaining RNA is precipitated from 70% alcohol (the supernatant contains the liberated base), redissolved in aqueous buffer, and treated with PME to remove the terminal phosphate.

If these operations can be carried out automatically in a completely enclosed modular system, it should be possible to maintain nuclease-free conditions and adjust the experimental parameters of each step for maximum efficiency. A simple block diagram illustrating this is presented in Fig. 6.2. The usual conception for this procedure is that the nucleic acid is first dephosphorylated with PME, and after a predetermined time the enzyme is inactivated by the addition of ethylenediamine tetraacetic acid (EDTA). The effect of the EDTA is to lower the pH and

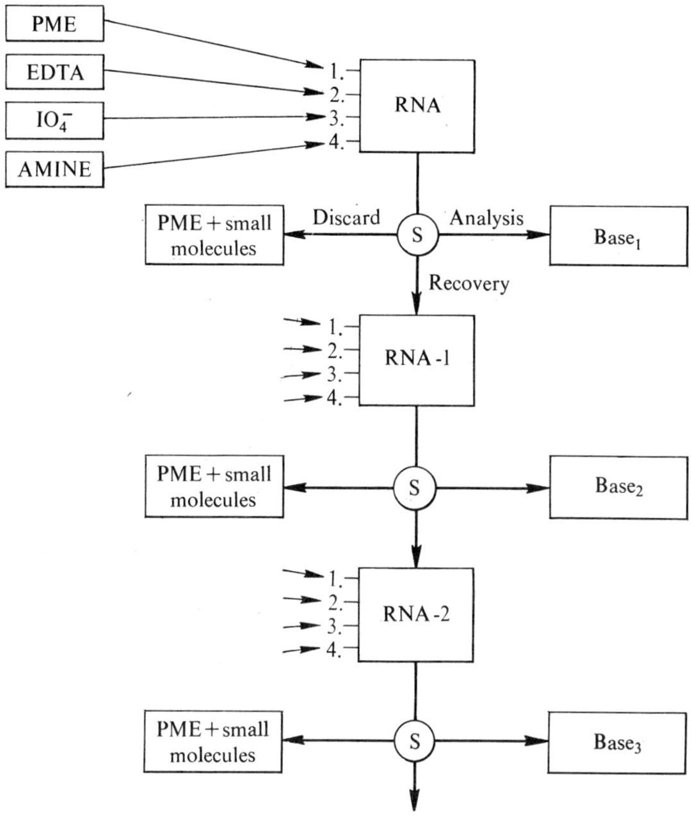

Fig. 6.2. Block diagram for the automatic, stepwise sequence analysis of RNA. The RNA enclosed in the box represents an RNA immoblized or bound to a solid support via its left terminal nucleotide. In succession, solutions of PME, EDTA, IO_4^-, and amine are passed through a column of bound RNA. Eluates are discarded or examined (where appropriate) for the presence of the liberated terminal base.

remove the zinc ions, which denature the PME and prevent recovery of activity at neutral pH. This is followed by the addition of periodate to the reaction mixture and, after a suitable incubation period, of an amine without the removal of the excess periodate. Under these simplified conditions, the cleavage step may result in a quantitative production of the free terminal base (Neu and Heppel, 1964).

Provided that the conditions can be adjusted to minimize nucleic acid chain breakage during these steps, the only problem occurs in the separation of the final reaction mixture into:

1. The remaining nucleic acid in a form ready for the next cycle.
2. The liberated base in a form that can be analyzed.
3. The molecules to be discarded (spent PME, periodate, EDTA, fragments from the cleavage reaction).

In the present state of the art, this may require a minimum of two separations such as an initial separation of the high molecular weight PME and RNA from the low molecular weight terminal base and other small molecules and fragments, followed by a separation of the PME from the RNA. Other steps may be needed to concentrate the RNA and adjust its buffered environment for the next cycle and to isolate the free base from the other small molecules for identification. In a completely enclosed, automatic system, these separation steps may require a number of different modules, such as chromatography columns and their associated elution solutions and control devices. If, for purposes of achieving optimum conditions, any of the first four steps must also be interrupted for purification or isolation of components, construction of an automatic sequenator would present a formidable engineering problem.

A simpler and possibly more satisfactory approach to automated sequence analysis would involve immobilization of the nucleic acid, e.g., in a solid phase, and flowing the appropriate solutions past it (see Weith and Gilham, 1969). Immobilization of a nucleic acid in a solid phase may be accomplished by entrapment within a molecular sieve material like cross-linked dextran (Yolles, 1967), or polyacrylamide (Mandeles, unpublished observations), or by chemical linkage to an insoluble support material like polystyrene (Letsinger and Mahadevan, 1965, 1966; Melby and Stroback, 1967; Hayatsu and Khorana, 1967). Assuming that an RNA has been attached to an insoluble solid support without any appreciable chain breakage, it may be used to fill a jacketed column, and each of the solutions in steps 1 through 4, and any others needed to attain optimum conditions, may be flowed through. Changes in reagents, pH, buffer, temperature, flow rate, etc., could be made very easily. The problem of separations would be reduced to a minimum and the cyclic process could be carried out with fewer detectors and control devices than when the nucleic acid is in solution.

With an automatic sequencer of this simple type, the efficiency of a set of experimental parameters could be checked with short polynucleotides or oligonucleotides of known sequence. For example, poly(alternating)AU could be used, provided that all of the chains in a preparation ended exclusively in A or in U at the right-hand terminus. If the chains ended in U, the odd-numbered cycles should produce only U's and the even-numbered cycles should produce only A's. Less than optimum conditions would be indicated by the extent of occurrence of A's in odd-numbered cycles and U's in even-numbered cycles. Experimental conditions could be altered until maximum synchrony was achieved.

STEPWISE DEGRADATION OF DNA

Much less attention has been given to the development of procedures for the chemical stepwise analysis of DNA. Proposals have been made by Reese *et al.* (1960) and Jones and Williamson (1960) that polydeoxyribonucleotides could be degraded by a base-catalyzed β-elimination reaction similar to that used with RNA. The procedure requires that the left terminal 5′-OH be converted into a suitable carbonyl function that will undergo a β-elimination. This reaction should proceed under conditions that will leave the rest of the nucleic acid intact. Specific methods for carrying out these reactions have been developed by Vizsolyi and Tener (1962) and are summarized in the scheme shown in Fig. 6.3. In the first step, the 5′-OH group at the left end of a DNA is oxidized to the corresponding uronic acid through use of a deactivated platinum catalyst and hydrogen peroxide. The deactivated catalyst is required to prevent irreversible adsorption of the DNA, and the peroxide promotes completion of the oxidation in a short time. Next, the uronic acid treated with dicyclohexylcarbodiimide (DCC) and then with propylamine is degraded, in the presence of 0.25 *N* NaOH, to a mixture containing the terminal base, *N*-propylfuramide, and the remainder of the polydeoxynucleotide chain. (When

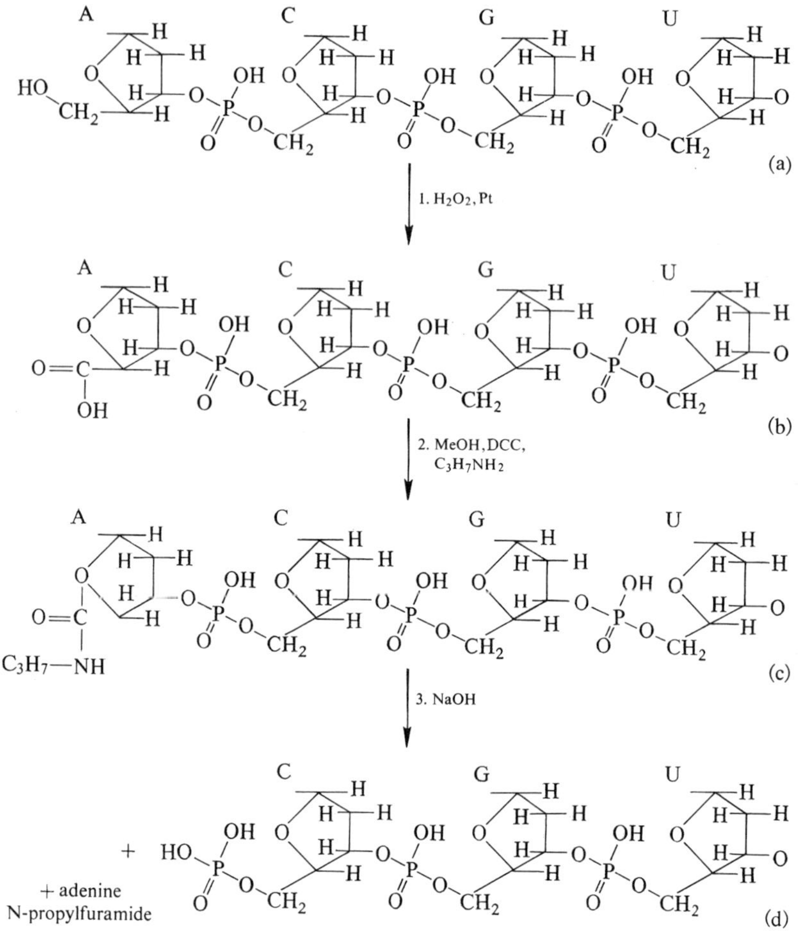

Fig. 6.3. Scheme for the stepwise degradation of DNA. In step 2, MeOH is methanol, DCC is dicyclohexyl carbodiimide.

this procedure was tried with thymidylyl-(3′ → 5′)thymidine, a fourth constituent, believed to be anhydrothymidine uronic acid, was found at the end.) The polydeoxyribonucleotide is then dephosphorylated with PME in preparation for the second elimination cycle. In principle, this stepwise degradation of DNA could also be automated and used as above in sequence analyses.

REFERENCES

Brown, D. M., Fried, M., and Todd, A. R. (1953). *Chem. Ind.*, 352.
Dutta, S. K., Jones, A. S., and Stacey, M. (1953). *Biochim. Biophys. Acta*, **10,** 613.
Hayatsu, H., and Khorana, H. G. (1967). *J. Am. Chem. Soc.*, **89,** 3880.
Holley, R. W., Apgar, J., and Merrill, S. H. (1961). *J. Biol. Chem.*, **236,** 42.
Jones, A. S., and Williamson, A. R. (1960). *Chem. Ind.*, 1624.
Khym, J. X., and Uziel, M. (1968). *Biochemistry*, **7,** 422.
Letsinger, R. L., and Mahadevan, V. (1965). *J. Am. Chem. Soc.*, **87,** 352b.
Letsinger, R. L., and Mahadevan, V. (1966). *J. Am. Chem. Soc.*, **88,** 5319.
Mandeles, S. (1967). *J. Biol. Chem.*, **242,** 3103.
Melby, L. R., and Stroback, D. R. (1967). *J. Am. Chem. Soc.*, **89,** 450.
Neu, H. D., and Heppel, L. A. (1964). *J. Biol. Chem.*, **739,** 2027.
Reese, C. B., Schofield, K., Shapiro, R., and Todd, A. R. (1960). *Proc. Chem. Soc.*, 290.
Steinschneider, A., and Fraenkel-Conrat, H. (1966). *Biochemistry*, **5,** 2735.
Uziel, M., and Gassen, H. G. (1969). *Fed. Proc.*, **28,** 409.
Uziel, M., and Khym, J. X. (1969). *Biochemistry*, **8,** 3254.
Vizsolyi, J. P., and Tener, G. M. (1962). *Chem. Ind.*, 263.
Weith, H. L., and Gilham, P. T. (1967). *J. Am. Chem. Soc.*, **89,** 5473.
Weith, H. L., and Gilham, P. T. (1969). *Science*, **166,** 1004.
Whitfeld, P. R. (1954). *Biochem. J.*, **58,** 390.
Whitfeld, P. R. (1965). *Biochim. Biophys. Acta*, **108,** 202.
Whitfeld, P. R., and Markham, R. (1953). *Nature*, **171,** 1151.
Yolles, R. (1967). *J. Am. Chem. Soc.*, **89,** 6344.

7.

END LABEL METHODS

A number of investigators have proposed to determine the sequence of bases near either end of nucleic acid molecules by chemical (or biochemical) modification of a particular end with an easily detected label. As usually formulated, the plan is to degrade the labeled nuclcic acid in some specific way and to find and determine the sequence of the oligonucleotide that is attached to the label. For the left end, the labels that have been suggested include ^{14}C-phosphoranilidates (Ralph *et al.*, 1962), ^{14}C-methyl pyrophosphate (Young and Khorana, 1963), ^{14}C-tritylation (Mandeles and Tinoco, 1963), ^{32}P (Richardson, 1965; Szekely and Sanger, 1970), and trisubstituted acetylation (Todd, 1965, personal communication). All of these labels may be attached specifically to the 5′ position at the left terminus. The labels for the right end usually require prior oxidation of the free 2′ and 3′ vicinal hydroxyls with periodate to form the dialdehyde as described in Chapter 6. This is followed by coupling with substances such as ^{35}S-thiosemicarbazide (Dulbecco and Smith, 1960), 2,4-dinitrophenylhydrazine (Monier *et al.*, 1960), a tetrazotized *o*-dianisidine derivative of 2-hydroxy-3-naphthoic acid hydrazide (Zamecnik, *et al.*, 1960; Stephenson and Zamecnik, 1961), polyacrylic acid hydrazide (Von Portatius *et al.*, 1961), aminoethyl cellulose (Zubay, 1962; Lee and Gilham, 1966), *p*-hydrazinobenzyl cellulose (Yolles, 1964), ^{3}H- or ^{14}C-isonicotinic acid hydrazide (Hunt, 1965), ^{14}C-semicarbazide (Steinschneider and Fraenkel-Conrat, 1966; Mandeles, 1967), and reduction of the dialdehyde to a dialcohol with ^{3}H-sodium borohydride (DeWachter and Fiers, 1967; Glitz *et al.*, 1968). These labels are selected for the stability of the product in contrast to the lability necessary in the β-elimination and cleavage reaction described in the preceding chapter.

Ideally, the sequence determination would be carried out as shown in the following scheme:

1. A hypothetical nucleic acid

- - - -GpGpUpGpApCpGpUpApUpGpCpCpCpA

is modified with a radioactive label, *R*, at the right end to form

- - - -GpGpUpGpApCpGpUpApUpGpCpCpCpA-*R*

which is then hydrolyzed with T_1 RNase to form oligonucleotides ending in Gp:

- - - -Gp + - - - -Gp + - - - -Gp + Gp + UpGp + ApCpGp + UpApUpGp + CpCpCpA-*R*

2. The products of the T_1 RNase hydrolysis are separated according to charge (phosphodiesters bear a single negative charge at neutral pH, phosphomonoesters are doubly charged) on a DEAE-cellulose (Tomlinson and Tener, 1963a,b) or DEAE-Sephadex column (Rushizky *et al.*, 1964) in the presence of 7 *M* urea to yield the pattern shown in Fig. 7.1. For this purpose, chromatography is ordinarily carried out at neutral pH with a linearly increasing gradient of chloride ions. The end oligonucleotide attached to the radioactive label, CpCpCpA-*R*, indicated by the shaded vertical bars, is shown as part of the second peak. In addition to the radioactive end oligonucleotide, this peak contains triply charged oligonucleotides of the general formula XpGp.

3. The peak containing the radioactive end oligonucleotide is end-treated with phosphomonoesterase to yield the mixture

XpG + CpCpCpA-*R* + Pi

It may be seen that by this stratagem (Tomlinson and Tener, 1963b) the net charges on all oligonucleotides, except CpCpCpA-*R*, from the second peak have been reduced by two. Chromatographic separation of this mixture according to charge should yield the pattern shown in Fig. 7.2. The radioactive end oligonucleotide may be isolated from this fractionation and its sequence determined by procedures described in

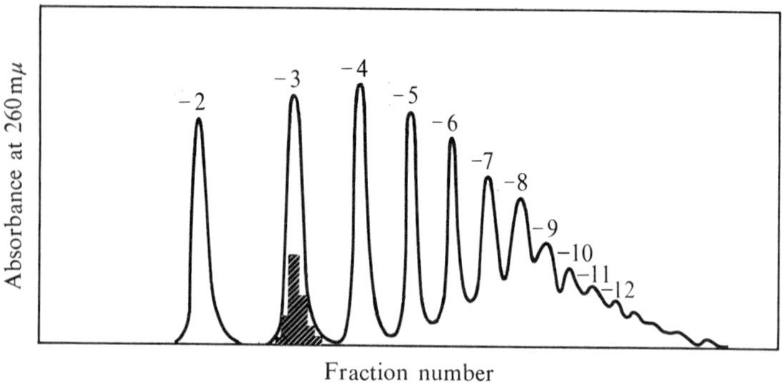

Fig. 7.1. Chain length separation pattern of oligonucleotides obtained from a hypothetical RNA. The right terminus is labeled with *R* and digestion is performed with T_1 RNase. The presence of *R* is indicated by the shaded bars.

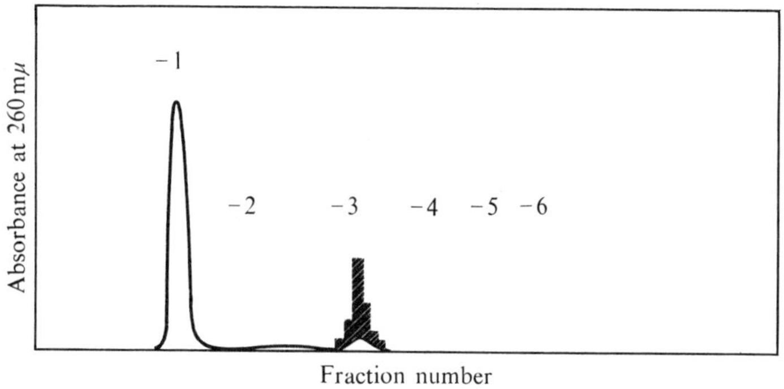

Fig. 7.2. Chain length separation of −3 peak of Fig. 7.1 after treatment with PME. The shaded bars indicate the presence of the label *R*.

the previous chapters. Because this tetranucleotide was obtained by T_1 RNase digestion of the hypothetical nucleic acid, the fifth position must be occupied by a Gp to yield GpCpCpCpA as the right end sequence. According to this deduction, the position of the first G residue on the right end of the molecule could be established by observing the pattern of radioactivity in the various peaks. For this example, the position of the G residue is given by $\varepsilon + 2$, where ε is the nominal charge of the oligonucleotides in the peak.

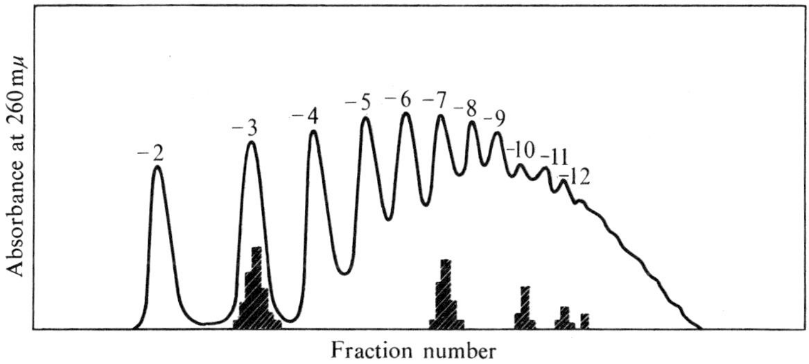

Fig. 7.3. Chain length separation pattern of oligonucleotides obtained from a hypothetical RNA after partial digestion with T_1 RNase. The RNA is labeled at the right terminus with *R* as in Fig. 7.1. The presence of *R* is indicated by shaded bars.

Under conditions where only partial hydrolysis of the nucleic acid by T_1 RNase occurs, chromatographic separation of the products would be expected to product the pattern shown in Fig. 7.3. Here the radioactivity is found in the peaks corresponding to −3, −7, −10, −12, and −13 charges. This indicates that a G is present in the 5th, 9th, 12th, 14th, and 15th positions. A similar procedure could be adopted for the products from a digestion of nucleic acid with pancreatic RNase. In this instance, the positions of the pyrimidines would be established.

In practice, slight variations from the expected fractionation patterns have been found (Mandeles, 1967; DeWachter and Fiers, 1967). This is illustrated in Fig. 7.4, where a comparison is made of the patterns obtained with a T_1 RNase hydrolysis of TMV RNA before and after the products are treated with PME. The dashed line shows the pattern with the untreated T_1 RNase digest. Essentially all of the oligonucleotides in this preparation are of the general structure $(Xp)_nGp$, where X = A, C, or U and $n = 0, 1, 2$, etc., and are terminated with the doubly charged phosphomonoester. The oligonucleotides in the PME-treated preparation are of the general structure $(Xp)_nG$, containing only singly charged phosphodiester bonds. The fractionation pattern of this preparation is shown as the solid line. It can be seen that oligonucleotides containing only phosphodiester linkages tend to be eluted earlier than those bearing the same number of charges but including one terminal phosphate: XpXpG is eluted before Gp,

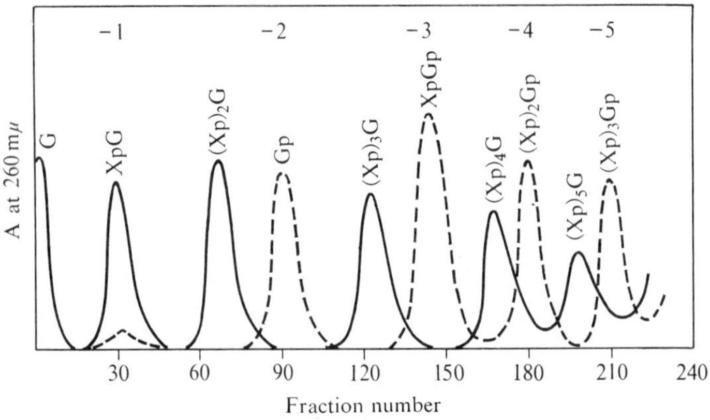

Fig. 7.4. Chain length separation patterns of oligonucleotides from T_1 RNase digests of TMV RNA before, - - - -, and after, ———, treatment with PME.

XpXpXpG is eluted before XpGp, and so on. From this result, it might be expected that the labeled end oligonucleotide from the hypothetical example above would be eluted in a position between the Gp and the XpGp peaks. Just such chromatographic behavior was observed in the isolation of the ^{14}C-semicarbazide-labeled end oligonucleotide from TMV RNA (Mandeles, 1967), the ^{3}H-labeled end oligonucleotide from bacteriophage MS2 RNA (DeWachter and Fiers, 1967), and f2 RNA (Glitz *et al.*, 1968). The end oligonucleotide from TMV RNA was found in a position between the Gp and XpGp peaks, indicating a structure of

XpXpXpX-*R*

The sequence of this oligomer was subsequently shown to be CpCpCpA. The end nucleotide from MS2 RNA was eluted in the hexa- and heptanucleotide regions, suggesting a chain length of nine bases: (Xp)$_8$X-*R*. The sequence of this oligonucleotide was subsequently shown to be UpUpApCpCpApCpCpCpA.

Labeled oligonucleotides from the left end of RNA may be expected to exhibit the same variation in chromatographic behavior. However, in extensive studies on the ^{32}P-labeled left terminal sequences of 16S and 23S RNA from a number of different species, Takanami (1967) and

Sugiura and Takanami (1967) have found pXp in the same peak as XpXpXp, pXpXp in the same peak as XpXpXpXp, and so on. The very slight shifts in position were attributed to differences in base composition.

SEQUENCE ANALYSIS OF LABELED OLIGONUCLEOTIDES FROM THE RIGHT TERMINUS

After chemical modification and isolation of an end oligonucleotide, the choice of a suitable sequence analysis procedure seems to depend as much on the personal preference of the investigator as on the chemical properties of the oligonucleotide. In the analysis of isonicotinic acid hydrazide-labeled oligonucleotides from *E. coli* ribosomal RNA, Hunt (1965) found evidence for two terminal sequences in 23S RNA but only one in 16S RNA. In this study the ribosomal RNA was separated by sedimentation in a sucrose density gradient and the fractions corresponding to 16S and 23S RNA were oxidized with periodate and coupled with ^{3}H-isonicotinic acid hydrazide (isoniazid). Only one equivalent of ^{3}H-isoniazid was incorporated per equivalent of RNA. This is in contrast to the two equivalents usually observed in the reaction of the dialdehyde with the substituted phenyl hydrazines or semicarbazide. The presumed monohydrazone structure shown in Fig. 7.5 is similar to that proposed by Khym (1963) for the reaction of the dialdehyde with amine.

Each of the isoniazid-labeled RNAs was digested in separate experiments with pancreatic and T_1 RNase and the products were fractionated by electrophoresis at low pH. The results provided evidence for the

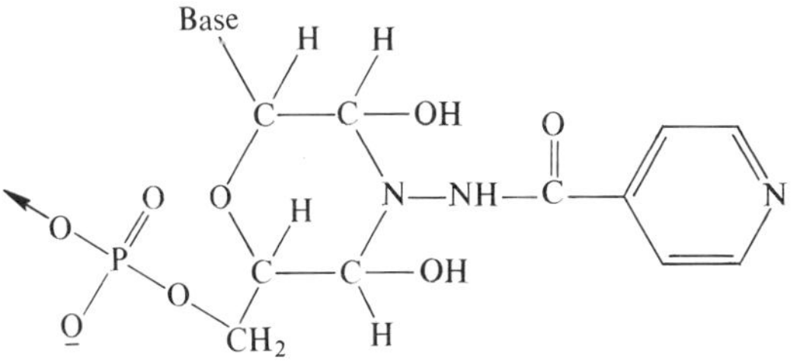

Fig. 7.5. Structure of isonicotinoyl monohydrazone.

sequences GpU and YpU for the right end of the 23S RNA, for YpA as the sequence on the right end of 16S RNA. It was apparent that further progress would require the use of additional procedures and other points of attack.

A somewhat different approach was adopted by Lee and Gilham (1966) and Weith and Gilham (1967) in the study of the right end oligonucleotide from a T_1 RNase digest of bacteriophage f2. They elected to treat the hydrolysate with periodate and then pass the mixture through a column of aminoethyl cellulose in the presence of 4 *M* NaCl at 4°. Under these conditions, the unoxidized internal fragments (with 3′-phosphomonoester ends) are eluted while the oxidized terminal fragment is retained. The chloride ions are then displaced by bicarbonate ions, and the end fragment is eluted from the column with a solution of *n*-propylamine at 45°. The oligonucleotide that is recovered is expected to lack its right terminal nucleoside because the conditions used for the recovery are similar to those used for the removal of the terminal base by β-elimination. As described in the previous chapter, the sequence of this end oligonucleotide was shown to be

-GpUpUpApCpCpApCpCpCpA

by correlation of data from base composition, pancreatic RNase hydrolysis, and chemical stepwise degradation.

The use of ^{14}C-semicarbazide provides an example of how the labeled end fragment may aid in the isolation of unlabeled end oligonucleotide (Mandeles, 1967). In this study, TMV RNA was oxidized with periodate and then allowed to react with ^{14}C-semicarbazide (Steinschneider and Fraenkel-Conrat, 1966) to form the semicarbazone derivative at the right end. Because at least two equivalents of semicarbazide reacted with each equivalent of TMV RNA, the bis-semicarbazone was presumably formed. The labeled TMV RNA was hydrolyzed with T_1 RNase and the products were chromatographed on DEAE-sephadex in the presence of 7 *M* urea. The radioactive semicarbazone-oligonucleotide was isolated and rechromatographed at low pH on AG1X2 to achieve a separation by base composition. By this time, the yield of labeled end oligonucleotide was too small to carry out any significant sequence analysis. More substantial amounts of end oligonucleotide were obtained, however, from unlabeled TMV RNA.

In this instance, a fragment was isolated from the same positions in the chain length fractionation as occupied by the labeled oligonucleotide. It was found to have the sequence

CpCpCpA

The identification here was unique in that it was made by a combination of absorption spectrophotometry (Mandeles and Cantor, 1966) and optical rotatory dispersion (Warshaw and Tinoco, 1965; Cantor and Tinoco, 1965) of the intact oligonucleotide. Confirmation of the sequence was obtained from the results of an SPDE hydrolysis of the oligonucleotide.

A label of possibly greater stability for the right end than the semicarbazone is afforded by the reduction of the periodate oxidized end with ^{3}H-sodium borohydride. This procedure was used by DeWachter and Fiers (1967) to study the right terminus of bacteriophage MS2 RNA, obtained from *E. coli* grown in ^{32}P, and labeled subsequently at the right end with ^{3}H. The presence of the two radioactive elements aided in subsequent determinations of molecular homogeneity. The MS2 RNA was then hydrolyzed with T_1 RNase and the products were separated according to charge on a column of DEAE-Sephadex at pH 7.6, in 7 *M* urea. As mentioned in a previous section, the ^{3}H-labeled end oligonucleotide was found in the region of hexa- or hepta-nucleotides (eight or nine negative charges).

The material in the radioactive peak was combined and rechromatographed at pH 3.5 on DEAE-cellulose to achieve a separation according to base composition. When the ^{3}H-labeled fractions from this chromatography were combined and the ratio of ^{32}P:^{3}H compared, it was estimated that the terminal oligonucleotide was still contaminated with about twice as much nonterminal material. This was followed by chromatography at pH 5.5. Here, the charge of the end oligonucleotide remains the same while the contaminating oligomers with terminal phosphates lose one charge. The contamination of the end oligonucleotide with nonterminal sequences at this stage was estimated to be near 12%. This was sufficient purity for experiments utilizing partial digestion with VPDE, but not for base composition and pancreatic RNase studies. For these, the end oligonucleotide was subjected to as many as five additional chromatographic purifications.

Table 7.1. Base composition of products from partial VPDE digestion of MS2 RNA right terminal oligonucleotide

Peak	Composition	Peak	Composition
1	UpN[a]	6	(2Cp,Ap,2Up)N
3	(Ap,2Up)N	7	(3Cp,2Ap,2Up)N
4	(Cp,Ap,2Up)N	8	(4Cp,2Ap,2Up)N
5	(2Cp,Ap,2Up)N		

[a] N symbolizes the nucleosides of A, C, or U obtained after alkaline hydrolysis.

Alkaline hydrolysis of the end oligonucleotide revealed the base composition $A_2C_5U_2$ with one of the A's identified as the right terminal nucleoside. Pancreatic hydrolysis showed that each of the internal A residues was followed by a C to yield the structure

(U,U,C,C,C,AC,AC,)A

This was followed by partial digestion with VPDE and separation of the products according to chain length essentially as described by Holley *et al.* (1964). Each of the fractions obtained was hydrolyzed with alkali and the base composition was determined. The results of this experiment are shown in Table 7.1.

Peak 1 established U as the left end of this terminal oligonucleotide. The relative position of the next two nucleotides shown in peak 3 may be deduced from the fact that no ApUp sequence was obtained from a pancreatic RNase digestion: the left end sequence must be UpUpAp. Peaks 4 to 8 establish the position of the next bases to yield

GpUpUpApCpCpApCpCpCpA

considering the fact that the terminal oligonucleotide was obtained originally from a T_1 RNase hydrolysis of MS2 RNA.

SEQUENCE ANALYSIS OF LABELED OLIGONUCLEOTIDES FROM THE LEFT TERMINUS

Of all the labels proposed for use at the left terminus of nucleic acid, the most extensive studies have been made with ^{32}P (Takanami, 1967; Sugiura and Takanami, 1967; Lane and Tamaoki, 1967). The procedure calls for the specific phosphoesterification of a free 5′-OH group as shown in the following scheme (Richardson, 1965; Novogrodsky and Hurwitz, 1965):

$$*pppA + HO\text{-}Np(Np)_nN\text{-}(OH)_2 \rightarrow ppA + *pNp(Np)_nN(OH)_2$$

This reaction is catalyzed by the enzyme polynucleotide kinase which is isolated from T_4 bacteriophage-infected *E. coli* cells. As indicated in the scheme, no perceptible phosphorylation of the free 2′- and 3′-OH groups occurs. In cases where the nucleic acid is naturally phosphorylated in the 5′-OH position at the left terminus, treatment with phosphomonoesterase genetrates the free 5′-OH group.

After esterification with ^{32}P, the nucleic acid is hydrolyzed separately with T_1 and pancreatic RNase, and the labeled left terminal oligonucleotide in each case is isolated essentially by the same chromatographic procedures used for the right terminal oligonucleotides. Examples of the chromatographic patterns obtained with T_1 and pancreatic RNase digests of 16S and 23S RNA from *E. coli* A19 are shown in Fig. 7.6 (Takanami, 1967). When labeled 23S RNA was digested with pancreatic RNase and separated on DEAE-Sephadex, almost all of the radioactivity emerged in the fractions corresponding to six negative charges, as shown in Fig. 7.6a. This means that the left end oligonucleotide has the structure

*pRpRpYp

When this labeled oligonucleotide was further digested with T_1 RNase and the products separated by chain length, the radioactivity was eluted in the fractions corresponding to four negative charges (Fig. 7.6b). From these data, the left end of 23S RNA is

pGpRpYp

In Fig. 7.6c, the chromatographic pattern from the separation of a T_1 RNase hydrolysate of labeled 16S RNA is shown. Here the radioactivity emerges in the eight negative charge peak. This indicates a left terminal structure of

*pNpNpNpNpGp

where N can be A, C, or U. Further hydrolysis of this end oligonucleotide with pancreatic RNase, followed by chain length chromatography, yielded radioactivity in the seven negative charge peak.

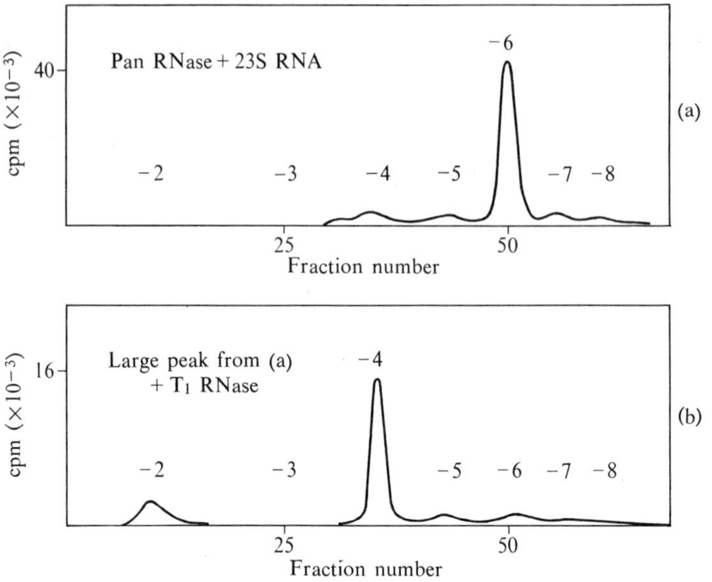

Fig. 7.6. Chain length chromatography of the left terminal oligonucleotide from *E. coli* A-19 rRNA. (a) 23S RNA labeled at left terminus with ^{32}P and digested with pancreatic RNase.

Apparently, the end oligonucleotide had been shortened by only one nucleotide; this meant that a pyrimidine preceded the right terminal G and that all of the other nucleotides must be purines (which are not guanines). The only structure that fits these data is

pApApApYpGp

In later studies on *E. coli* rRNA, Sugiura and Takanami (1967) showed that the sequence at the left end of 16S RNA was in fact pApApAp-UpGp, and that of 23S RNA was pGpGpUp.

The left terminal sequences of rRNA from a number of other organisms were determined in the same fashion. These sequences are summarized in Table 7.2. It may be seen that two structures were determined for the left end of *S. cerevisiae* 23S RNA, indicating the presence of two molecular species. The same sort of heterogeneity may exist in the rRNA from the other species in view of the fact that as much as 45% of the radioactivity in the ^{32}P end label was not found in

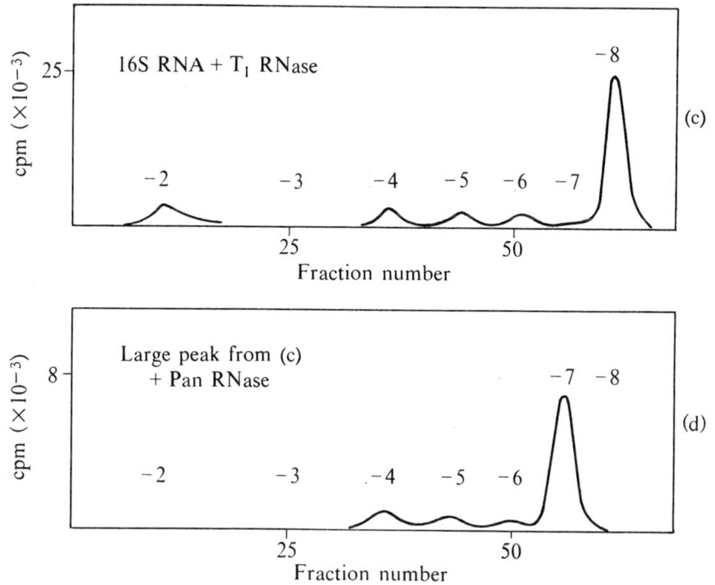

(b) −6 peak from (a) digested with T_1 RNase. (c) 16S RNA labeled at left terminus with ^{32}P and digested with T_1 RNase. (d) −8 peak from (c) digested with pancreatic RNase.

the structures listed in Table 7.2, but in other as yet undetermined oligonucleotides.

Double-Label Method. It would appear that the application of end label methods is effective in the determination of a small number of bases (in some cases as many as 10 or 11) near the termini of nucleic acids. Extension of the method to the determination of adjacent interior sequences seems to be limited by the difficulty of isolating and identifying these interior sequences from populations of heterogeneous oligonucleotides. This may be accomplished however, by application of a double-label method (Mandeles and Tinoco, 1963) which is a logical extension of the end label procedures.

Briefly, the process involves the labeling of one end of a nucleic acid, followed by partial enzymatic digestion to form an oligonucleotide mixture as indicated in Fig. 7.3. After separation of the labeled oligonucleotides from the unlabeled oligonucleotides, and from each other

Table 7.2. Left terminal sequences of ribosomal RNA[a]

Organism	23S	16S
E. coli	pG-G-U-	pA-A-A-U-G-
Bacillus cereus	pU-X-X-G-	pU-X-X-X-X-G-
B. subtilis	pU-X-X-X-G-	pU-X-X-X-X-G-
B. stearothermophilis	pU-X-X-X-G-	pU-X-X-X-X-G-
Sarcina lutea	pA-A-G-Py-	pU-X-X-X-X-G-
Saccharomyces	pU-U-G-	pUpX-X-X-X-G-
Saccharomyces cerevisiae	pA-A-A-C-	

[a] Adapted from Takanami (1967) and Sugiura and Takanami (1967).

according to chain length, a second label is attached to the other end. The double-labeled oligonucleotide is digested to completion with the same enzyme used in the partial hydrolysis, and the oligonucleotide bearing the second label is isolated and its sequence is determined. In this manner, the original nucleic acid sequence can be reassembled in terms of sequential oligomeric fragments. The features of this general method include: construction of long sequences from shorter fragments, unequivocal positioning of fragments, and independence from quantitative procedures. For clarity, this procedure is presented symbolically in the outline in Fig. 7.7.

A simpler and possibly more efficient approach is afforded by a modification suggested by Tinoco (personal communication, 1967). In this procedure, nucleic acid is labeled with a 5′-phosphomonoester at the left end through use of ^{32}P-ATP and polynucleotide kinase. The labeled polynucleotide is digested partially with T_1 RNase, which is then removed by bentonite-phenol extraction or pronase action or both. The mixture of partially digested labeled oligonucleotides is then hydrolyzed to completion with SPDE, which acts only on ends bearing a free 5′-OH group. As a result, only oligonucleotides bearing labeled 5′-phosphate groups are resistant to hydrolysis; the others are converted into nucleoside-3′-phosphates. The procedure up to this point is shown schematically in Fig. 7.8. The nucleotide sequence 3 on p. 73 may be selected as an example. It is first labeled with a radioactive phosphate in the 5′-hydroxyl position at the left end and then treated with the two hydrolytic enzymes. The products, when separated according to charge, should yield the chromatographic pattern shown in Fig. 7.9. The bulk of the material is eluted as monomer with two negative

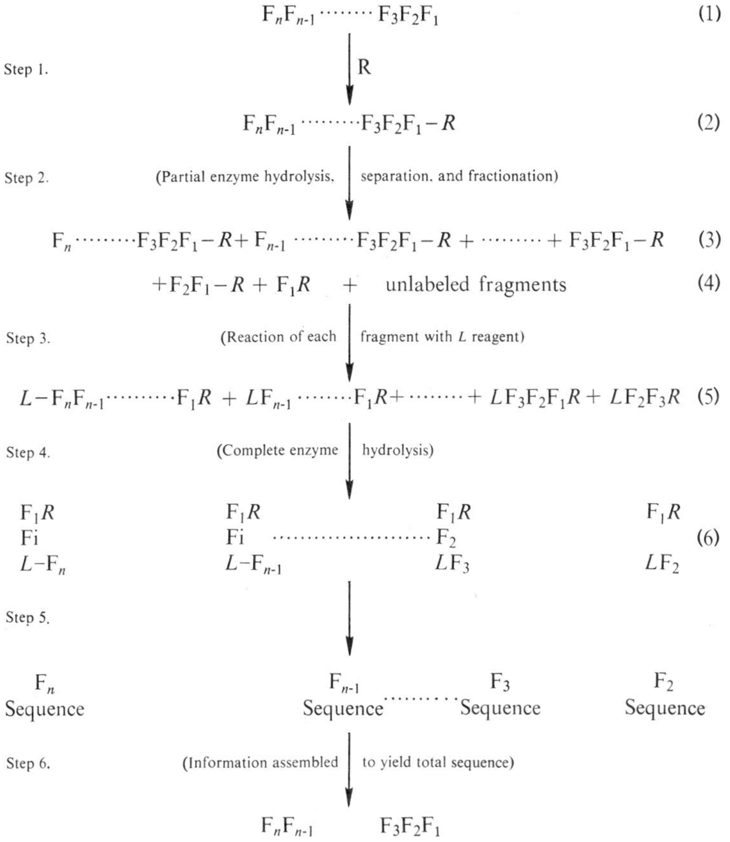

Fig. 7.7. Schematic of a double-label method. $F_nF_{n-1} \ldots F_3F_2F_1$ is the RNA represented as a structure of consecutive, specific fragments which depends on the specific nuclease employed. R and L represent, respectively, the right and left terminal labels.

charges in the first peak, while the next smaller peaks represent oligonucleotides with ^{32}P attached to the left end. As indicated in this figure, none of these later peaks is contaminated with unlabeled oligonucleotides owing to exonucleolytic action on the unprotected ends of the unlabeled oligonucleotides. This should simplify the problem of removing labeled fragments from unlabeled fragments as required in step 2 of Fig. 7.7, and is analogous to the use of aminoethyl

GpUpApGpCpGpCpCpGpApUpGpGpUpApGp

↓ $AT^{32}p$ + polynucleotide kinase

*pGpUpApGpCpGpCpCpGpApUpGpGpUpApGp

↓ Partial T_1 RNase digestion, SPDE complete hydrolysis

*pGp +
*pGpUpApGp +
*pGpUpApGpCpGp +
*pGpUpApGpCpGpCpCpGp +
*pGpUpApGpCpGpCpCpGpApUpGp +
*pGpUpApGpCpGpCpCpGpApUpGpGp +
*pGpUpApGpCpGpCpCpGpApUpGpGpUpApGp +
mononucleoside – 3′-phosphates

Fig. 7.8. Schematic of sequence analysis of hypothetical oligonucleotide using combined partial T_1 RNase and complete SPDE digestions.

cellulose, polyacrylic acid hydrazide, and *p*-hydrazinobenzyl cellulose for right end labels. This procedure has an additional advantage in that many nucleic acids, such as ribosomal RNA (Takanami, 1967), f2 RNA (Takanami, 1966), R17 RNA (Roblin, 1968), satellite tobacco necrosis virus (STNV) RNA (Wimmer and Reichmann, 1968),* and MS2 RNA (Glitz *et al.*, 1968),† occur naturally with phosphate or pyrophosphate esters in the 5′-hydroxyl position at the left end. Nucleic acids such as these could be used for sequence studies with no prior treatment other than purification, provided that sufficient amounts could be obtained for fractionation and analysis.

* The structure of the left terminal oligonucleotide from a pancreatic RNase hydrolysis was, in fact, shown to be ppApGpUp.

† The structure of the left terminal oligonucleotide from a pancreatic RNase hydrolysis was determined as pppGpGpUp.

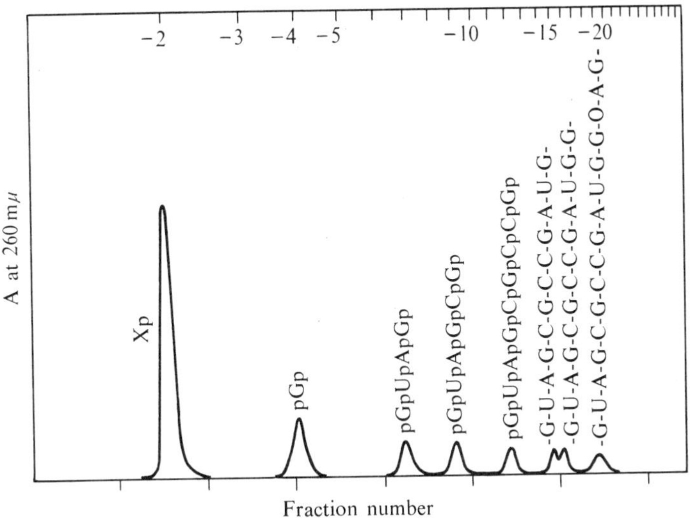

Fig. 7.9. Chain length separation of products of combined T_1 RNase and SPDE action on hypothetical oligonucleotide shown in Fig. 7.8.

APPLICATION TO OLIGONUCLEOTIDE SEQUENCE ANALYSIS

Up to now, end label procedures have been discussed in reference to the determination of sequences in RNA of high molecular weight. It is obvious, from the oligonucleotide used as an illustration in Fig. 7.8, that end label methods may be used for shorter oligonucleotides (in the range 12 to 100) as well. The combination of end label techniques and the fingerprint methods (Brownlee *et al.*, 1968; Szekely and Sanger, 1970) may form the basis of the rapid, efficient sequence analysis procedure needed for the longer oligonucleotides. Given such a combination, the plan would be to determine the sequence of the smallest oligomer from digestion products, as shown in Figs. 7.8 and 7.9. The next larger oligomer would contain this first sequence plus the additional sequence up to the second Gp. The additional sequence could be identified in a fingerprint following complete T_1 RNase digestion of the second peak. In a similar manner, the next contiguous sequence would be obtained from the third peak, and so on.

One limitation that this method shares with the more general double-label method is the difficulty of separating oligonucleotides of chain length *n* from those of chain length $n + 1$. As indicated in Fig. 7.9, separations of oligonucleotides with 15 negative charges from those with 16 negative charges are not complete, as a practical matter. The exploitation of the more subtle differences, such as base composition or sequence, may also be ineffective when adjacent peaks are very similar in these respects.

A second factor that tends to blur the otherwise clean fractionations is the presence of contaminating activities in the various enzyme preparations. In common with other sequence analysis procedures, enzymes with a high degree of purity and specificity are required.

REFERENCES

Brownlee, G. G., Sanger, F., and Barrell, B. G. (1968). *J. Mol. Biol.*, **34**, 379.

Cantor, C. R., and Tinoco, I., Jr. (1965). *J. Mol. Biol.*, **13**, 65.

DeWachter, F., and Fiers, W. (1967). *J. Mol. Biol.*, **30**, 507.

Dulbecco, R., and Smith, J. D. (1960). *Biochim. Biophys. Acta*, **39**, 361.

Glitz, D. G., Bradley, A., and Fraenkel-Conrat, H. (1968). *Biochim. Biophys. Acta*, **161**, 1.

Holley, R. W., Madison, J. T., and Zamir, A. (1964). *Biochem. Biophys. Res. Commun.*, **17**, 389.

Hunt, J. A. (1965). *Biochem. J.*, **95**, 541.

Khym, J. X. (1963). *Biochemistry*, **2**, 344.

Lane, B. G., and Tamaoki, T. (1967). *J. Mol. Biol.*, **27**, 335.

Lee, J. C., and Gilham, P. T. (1966). *J. Am. Chem. Soc.*, **88**, 5685.

Mandeles, S. (1967). *J. Biol. Chem.*, **242**, 3103.

Mandeles, S., and Cantor, C. R. (1966). *Biopolymers*, **4**, 759.

Mandeles, S., and Tinoco, I., Jr. (1963). *Biopolymers*, **1**, 183.

Monier, R. M., Stephenson, M. L., and Zamecnik, P. C. (1960). *Biochim. Biophys. Acta*, **43**, 1.

Novogrodsky, H., and Hurwitz, J. (1965). *Fed. Proc.*, **24**, 602.

Ralph, R. K., Young, R. J., and Khorana, H. G. (1962). *J. Am. Chem. Soc.*, **84**, 1490

Richardson, C. (1965). *Proc. Natl. Acad. Sci.*, **54**, 158.

Roblin, R. (1968). *J. Mol. Biol.*, **31**, 51.

Rushizky, G. W., Bartos, E. M., and Sober, H. A. (1964). *Biochemistry*, **3**, 626.

Steinschneider, A., and Fraenkel-Conrat, H. (1966). *Biochemistry*, **5**, 2735.

Stephenson, M. L., and Zamecnik, P. C. (1961). *Proc. Natl. Acad. Sci.*, **47**, 1627.

Sugiura, M., and Takanami, M. (1967). *Proc. Natl. Acad. Sci.*, **58**, 1595.

Szekely, M., and Sanger, F. (1970). *J. Mol. Biol.*, **43**, 607.

Takanami, M. (1966). *Cold Spring Harbor Symp. Quant. Biol.*, **31**, 611.

Takanami, M. (1967). *J. Mol. Biol.*, **23**, 135.

Todd, A. (1965). Personal communication.

Tomlinson, R. V., and Tener, G. M. (1963a). *Biochemistry*, **2**, 697.
Tomlinson, R. V., and Tener, G. M. (1963b). *Biochemistry*, **2**, 703.
von Portatius, H., Doty, P., and Stephenson, M. L. (1961). *J. Am. Chem. Soc.*, **83**, 3351.
Warshaw, M. M., and Tinoco, I., Jr. (1965). *J. Mol. Biol.*, **13**, 54.
Weith, H. L., and Gilham, P. T. (1967). *J. Am. Chem. Soc.*, **89**, 5473.
Wimmer, E., and Reichmann, M. E. (1968). *Science*, **160**, 1452.
Yolles, R. (1964). *Biochim. Biophys. Acta*, **87**, 583.
Young, R. J., and Khorana, H. G. (1963). *J. Am. Chem. Soc.*, **85**, 244.
Zamecnik, P. C., Stephenson, M. L., and Scott, J. F. (1960). *Proc. Natl. Acad. Sci.*, **47**, 1627.
Zubay, G. (1962). *J. Mol. Biol.*, **4**, 347.

8.

SEQUENCE ANALYSIS OF HIGH MOLECULAR WEIGHT RNA

The main difficulty in determining base sequences in ribosomal, messenger, and viral RNA stems from the large size of these molecules. They range in chain length from about 1300 bases in satellite tobacco necrosis virus RNA and 1500 bases in 16S ribosomal RNA to 6000 or more bases in viral RNA. When contrasted with the approximately 75 bases of tRNA and the 120 bases of 5S RNA, the problems of determining sequences in the larger molecules would seem to be at least ten- to twenty-fold greater than those encountered with the smaller molecules. Even with the most successful method to date, the overlap method, the assembly of the necessarily completely accurate oligonucleotide catalogs from specific nuclease digests of high molecular weight RNA is a monumental task.

The chances would be increased for a total determination of sequences in these larger molecules if they could be divided into smaller subunits of up to 200 bases or if the analysis could be limited to certain specific segments. These approaches as well as a variety of others have been suggested to cope with the problems of sequence analysis of high molecular weight nucleic acids, and are described in this chapter.

FORMATION OF SUBUNITS

One of the most favored approaches to the formation of subunits of RNA depends on the fact that certain nucleic acids tend to form stable configurations in solution. In the case of tRNA, this stable configuration is believed to be the familiar cloverlike conformation with a major portion of the bases paired with their complements to form short helical regions (see Chapter 12). Under the conditions of low temperature and 0.1 *M* salt which favor the formation of these helical regions, much of the nucleic acid will be resistant to nuclease action. Conversely, the portions of nucleic acid not involved in base pairing will be susceptible to nuclease action, thus permitting a limited, specific digestion of the tRNA.

tRNA SUBUNITS

Because of this situation, Penswick and Holley (1965) were able to cleave alanyl-tRNA with T_1 RNase at precisely one point as shown in Fig. 8.1. It can be seen that, when the sequence of alanyl-tRNA is arranged so that there is maximum complementarity between the various portions of the molecule, the three open loops contain unpaired guanine bases. If the molecule is treated with T_1 RNase at low temperature for a brief enough time, only the phosphate bonds at these points should be hydrolyzed. When the reaction was carried out under these conditions (4 minutes at 0° in 0.02 *M* $MgCl_2$), only two fragments were produced and they were nearly equal in length. This indicates that the G bond in the middle loop is more susceptible to T_1 RNase digestion than are the G bonds in the other two loops, and that slightly longer hydrolysis might produce cleavages at these other exposed points. This is in fact what was observed by Apgar *et al.* (1965); hence alanyl-tRNA could be divided roughly into either two or four subunits, depending on the conditions of hydrolysis.

SUBUNITS FROM HIGH MOLECULAR WEIGHT RNA

The possibility that a natural, high molecular weight nucleic acid assumes a specific configuration in solution is raised when one considers that the RNA of icosahedral viruses and of ribosomes is probably folded in some specific way to permit efficient packing in the former and correct functioning in the latter. It may be assumed that the specific folding is accomplished to a large extent by predetermined

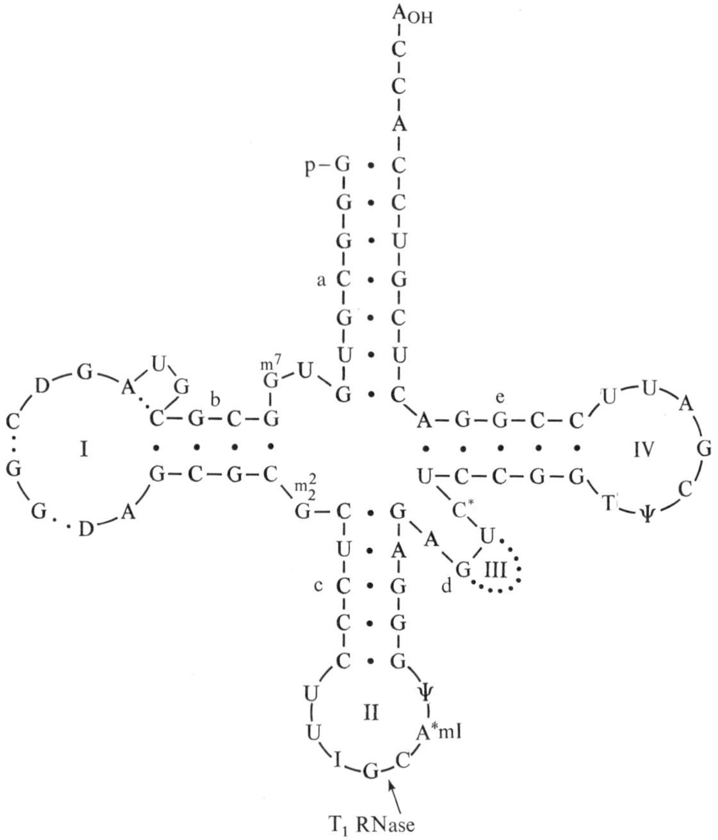

Fig. 8.1. Formation of subunits from alanyl-tRNA. Under conditions of 0°, 4 minutes, and 0.02 M $MgCl_2$, alanyl-tRNA is cleaved by T_1 RNase mainly at point shown.

regions of helical, complementary sequences and that RNA in solution will form roughly the same helical regions as the RNA *in situ* (see Cox, 1969a, for review). As a result, fewer portions of the molecule would be sensitive to nuclease action and a limited digestion should produce a small number of well-defined fragments.

Evidence in support of this idea has been obtained from the limited digestion of rabbit reticulocyte rRNA and μ2 virus RNA with T_1 RNase (Gould *et al.*, 1969), of reticulocyte rRNA with pancreatic RNase

(Gould, 1966), of yeast and *E. coli* rRNA with T_1, and pancreatic RNases (McPhie *et al.*, 1966) of Qβ RNA with pancreatic RNA with pancreatic RNase (Bassel and Spiegelman, 1967), and of R17 RNA with pancreatic RNase (Thach and Baedtker, 1969), T_1 RNase (Adams *et al.*, 1969; Nichols 1970), and RNase IV (Spahr and Gesteland, 1968). In a typical study (Gould *et al.*, 1969) rabbit reticulocyte 19S and 30S rRNAs were digested with T_1 RNase at 22° for short periods of time. The substrate to enzyme ratios were in excess of 20:1 (weight: weight), and separation of the fragments in the hydrolysate was achieved with polyacrylamide gel electrophoresis techniques. Over a third of the total material from either the 19S RNA or the 30S RNA was recovered in a small number of bands of comparatively high molecular weight. From the position of these bands in the gel, it was estimated that the molecular weights ranged from 32,000 to 150,000 for the six bands from the 19S RNA digest. The molar yield of most of these components was 1.0 or less, indicating the possibility that each band contained only a single molecular species.

In the limited hydrolysis of μ2 virus RNA, 60 percent of the total RNA was found in 14 discrete bands whose positions indicated a molecular weight range of 34,000 to 220,000. The molar yield for each of these components was less than 1.0, providing evidence for the molecular homogeneity of the fraction. On the basis that there are approximately three nucleotides for every 1000 molecular weight units, it may be calculated that all except the very largest fragments are small enough for sequence analysis by conventional methods.

SUBUNITS FROM INTACT RIBOSOMES

The only major problem in the preceding approach to the formation of subunits may be the ability to obtain the same fragments reproducibly from experiment to experiment. In the case of rRNA, greater assurance about the reproducible fragmentation is afforded by the studies of the effect of nuclease on the intact ribosome (Cox, 1969). It was found that treatment of rabbit reticulocyte ribosomes with pancreatic RNase resulted in the loss of 20 to 30 percent of the ribosomal RNA without appreciably altering the sedimentation coefficient or appearance of the ribosomes in electron photomicrographs. The sizes of the RNA fragments isolated from the RNA depleted ribosomes indicated that there are about 40 nuclease-sensitive sites in the larger subparticle and 20 in the smaller. It was estimated that, on the average, 105 nucleotides separated these susceptible sequences.

These results are consistent with a model of ribosomal structure which requires nucleotide sequence-dependent interactions between the ribosomal proteins and the RNA (Cox and Bonanou, 1969). According to the model, a small number of single-strand RNA loops are exposed on the surface of the ribosome. The positions of these exposed loops are determined by specific helical regions in other parts of the RNA as well as by specific interaction with ribosomal protein. A diagram of the proposed model is shown in Fig. 8.2. Assuming that

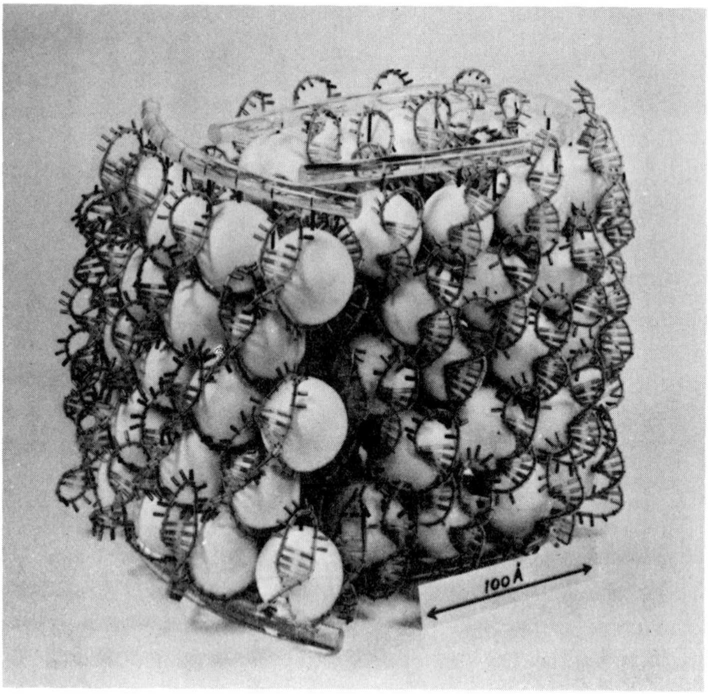

Fig. 8.2. Model of reticulocyte ribosome (from Cox and Bonanou, 1969). Unpaired portions of the RNA may be seen exposed on the surface.

the same structure is present in all ribosomes of the same species, complete digestion of the exposed nucleic acid should result in the reproducible formation of specific subunits.

CHARACTERIZATION OF A LONG FRAGMENT FROM R17 RNA

Using the conditions for the limited digestion of R17 RNA, Adams *et al.* (1969) have isolated a specific fragment that consisted of 57 nucleotides. They have determined the sequence of this fragment and have demonstrated that it contains the correct order of codons for a portion of the virus coat protein. This is the first demonstration of the colinearity between an amino acid sequence and its corresponding messenger.

The isolation of the fragment depended on the fact that only a small number of unique, long-chain oligonucleotides can be expected from a T_1 RNase hydrolysis of the R17 coat protein cistron. This is illustrated in Fig. 8.3, which shows the amino acid sequence in the coat protein (Weber, 1967) together with the corresponding nucleotide sequence as specified by the generalized genetic code shown in Table 8.1. In Fig. 8.3, Y stands for pyrimide; R for purine; 3 for A, U, or C; and N for any of the four bases. The sequence of nucleotides corresponding to the amino acid in the coat protein may be written with the "unlucky" choice of G taking the place of every R and N, as would be permitted with a degenerate code. Under these conditions, the longest sequence of bases devoid of guanine would occur between base number 7 and base number 15, corresponding to amino acids 3 to 5. This is shown here in a portion of the sequence in Fig. 8.3, where G substitutes for R and N:

Ala Ser Asn Phe Thr Gln Phe Val
GCG UCG <u>AAY UUY ACG CAG</u>

The underlined sequence is the longest oligonucleotide that would be produced from a T_1 RNase digest of the coat protein cistron when the "unlucky" substitutions are made. This hypothetical 9-mer could be expanded to a 10-mer if the codon AGY were used for Ser rather than UCG. In this instance, the 10-mer should be

Y AAY UUY ACG

Amino Acid No.					5					10					15
1	Ala	Ser	Asn	Phe	Thr	Gln	Phe	Val	Leu	Val	Asn	Asp	Gly	Gly	Thr
	GCN	UCN	CAR	UUY	ACN	AAY	UUY	GUN	CUN	GUN	CAR	GAR	GGN	GGN	ACN
		AGY							UUR						
16	Gly	Asn	Val	Thr	Val	Ala	Pro	Ser	Asn	Phe	Ala	Asn	Gly	Val	Ala
	GGN	CAR	GUN	ACN	GUN	GCN	CCN	UCN	CAR	UUY	GCN	CAR	GGN	GUN	GCN
								AGY							
31	Glu	Trp	Ile	Ser	Ser	Asn	Ser	Arg	Ser	Gln	Ala	Tyr	Lys	Val	Thr
	GAY	UGG	AU3	UCN	UCN	CAR	UCN	CGN	UCN	AAY	GCN	UAY	AAR	GUN	ACN
				AGY	AGY		AGY	AGR	AGY						
46	Cys	Ser	Val	Arg	Gln	Ser	Ser	Ala	Gln	Asn	Arg	Lys	Tyr	Thr	Ile
	UGY	UCN	GUN	CGN	AAY	UCN	UCN	GCN	AAY	CAR	CGN	AAR	UAY	ACN	AU3
		AGY		AGR		AGY	AGY				AGR				
61	Lys	Val	Glu	Val	Pro	Lys	Val	Ala	Thr	Gln	Thr	Val	Gly	Gly	Val
	AAR	GUN	GAY	GUN	CCN	AAR	GUN	GCN	ACN	AAY	ACN	GUN	GGN	GGN	GUN
76	Glu	Leu	Pro	Val	Ala	Ala	Trp	Arg	Ser	Tyr	Leu	Asn	Met	Glu	Leu
	GAY	CUN	CCN	GUN	GCN	GCN	UGG	CGN	UCN	UAY	CUN	CAR	AUG	GAY	CUN
		UUR						AGR	AGY		UUR				UUR
91	Thr	Ile	Pro	Ile	Phe	Ala	Thr	Asn	Ser	Asp	Cys	Glu	Leu	Ile	Val
	ACN	AU3	CCN	AU3	UUY	GCN	ACN	CAR	UCN	GAR	UGY	GAY	CUN	AU3	GUN
									AGY				UUR		
106	Lys	Ala	Met	Gln	Gly	Leu	Leu	Lys	Asp	Gly	Asn	Pro	Ile	Pro	Ser
	AAR	GCN	AUG	AAY	GGN	CUN	CUN	AAR	GAR	GGN	CAR	CCN	AU3	CCN	UCN
						UUR	UUR								AGY
121	Ala	Ile	Ala	Ala	Asn	Ser	Gly	Ile	Tyr						
	GCN	AU3	GCN	GCN	CAR	UCN	GGN	AU3	UAY						
						AGY									

Fig. 8.3. R17 coat protein sequences with corresponding nucleic acid codons.

On the other hand, when the lucky substitutions are made, more than 17 sequences of bases can be found that have a single terminal G and contain 9 or more bases. In these instances, N is substituted only by A, U, or C, and R by A. The sequences are shown in Table 8.2. It can

Table 8.1. Genetic code

Amino Acid	Codon	Amino Acid	Codon	Amino Acid	Codon
Ala	GCN	Gly	GGN	Pro	CCN
Arg	CGN, AGR	His	CAY	Ser	UCN, AGY
Asn	AAY	Ile	AU3	Thr	ACN
Asp	GAY	Leu	UUR, CUN	Try	UGG
Cys	UGY	Lys	AAR	Tyr	UAY
Gln	CAR	Met	AUG	Val	GUN
Glu	GAR	Phe	UUY	Stop	UGA, UAR

Table 8.2. Hypothetical sequences from T_1 RNase hydrolysis of R17 coat protein cistron

									Chain Length
1.	CN	UCN	AAY	UUY	ACN	CAR	UUY	G	21
		AGY							
2.	CN	CCN	UCN	AAY	UUY	G			15
			AGY						
3.	AU3	UCN	UCN	AAY	UCN	CG			17
		AGY	AGY						
4.	CN	UAY	AAR	G					9
5.	N	CAR	UCN	UCN	G				11
			AGY	AGY					
6.	CN	CAR	AAY	CG					10
7.	N	AAR	UAY	ACN	AU3	AAR	G		17
8.	UN	CCN	AAR	G					9
9.	CN	ACN	CAR	ACN	G				12
10.	AR	UUR	CCN	G					9
		CUN							
11.	N	UCN	UAY	UUR	AAY	AUG			16
		AGY		CUN					
12.	AR	UUR	ACN	AU3	CCN	AU3	UUY	G	21
		CUN							
13.	CN	ACN	AAY	UCN	G				12
14.	AR	UUR	AU3	G					9
		CUN							
15.	N	UUR	UUR	AAR	G				11
		CUN	CUN						
16.	N	AAY	CCN	AU3	CCN	UCN	G		17
						AGY			
17.	CN	AAY	UCN	G					9
			AGY						

be seen that two of the hypothetical sequences contain 21 bases, three contain 17, and two others contain 16 and 15, respectively. This tally neglects the ways in which the longer oligonucleotides such as the 21-mers can be shortened and still contain more than 15 bases. If any of these hypothetical oligonucleotides actually occurs in a T_1 RNase

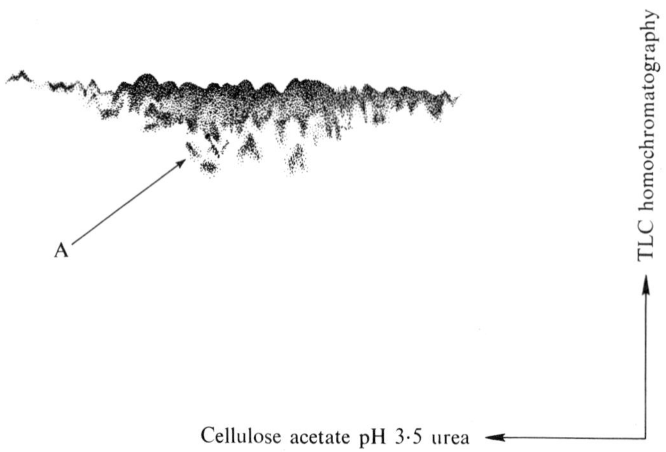

Fig. 8.4. Homochromatographic fractionation of a T_1 RNase digest of R17 RNA (from Adams *et al.*, 1969). Fragment A, indicated by the arrow, contained the codons corresponding to a portion of the coat protein.

digest of R17 RNA, the chances are good that it will be unique and easily separated from the other oligonucleotides.

In experimental terms, one could carry out a T_1 RNase hydrolysis of R17 RNA and focus one's attention on separating the longest oligomers from each other. Each long oligonucleotide could then be isolated and analyzed for chain length and base composition, and these values compared with the values expected from the oligonucleotides shown in Table 8.2.

This, in fact, was carried out by Adams *et al.* (1969). A T_1 RNase digest was made of ^{32}P-R17 RNA and the products were separated by a combination of ionophoresis on cellulose acetate and homochromatography on thin-layer DEAE-cellulose (Brownlee and Sanger, 1969). An example of this separation is shown in Fig. 8.4. Here most of the small oligonucleotides migrate rapidly and appear as the diffused front, while some of the longer oligonucleotides move more slowly and are easily seen as well-resolved spots in the radioautogram. One of these spots, labeled A in the fingerprint, was hydrolyzed with alkali and yielded the base composition (after rounding the values) $A_7C_4U_9G$. This

Table 8.3. Sequence analysis of oligonucleotide A

C List[a]	U List[b]
1. C	1. AA
2. G	2. (C,U)A
3. (AU,U_2)C	3. UUAA
4. (AAU,U_3)C	4. (U_2,C_2)AA
5. (AAU,U)AAC	5. (U_4,C)G

[a] Pancreatic RNase products from CMCT derivative.

[b] U_2 RNase products, except as noted in text.

indicated that, of all the long-oligonucleotide possibilities shown in Table 8.2, only the two 21-mers need be considered.

A pancreatic RNase digest of fragment A produced the following shorter oligonucleotides:

1 G, 3 C, 6 U, 1 AU, 2 AAU, and 1 AAC

This result indicated that fragment A could be identified with the second 21-mer (oligonucleotide 12 in Table 8.2) because there is no way that the data can be made to fit the general sequence of the first 21-mer (oligonucleotide 1). The remaining candidate provided two alternative sequences:

AR UUR ACN AU3 CCN AU3 UUY G and
AR CUN ACN AU3 CCN AU3 UUY G

Here, again, the data can be made to fit only one of the sequences; the upper one.

In order to complete the sequence analysis of fragment A, additional data were provided by the use of the CMCT blocking agent to produce oligonucleotides ending in C (discussed in Chapter 4; see Chapter 11 for procedure) and the enzyme U_2 RNase to produce oligonucleotides that end in A (see Chapter 11 for conditions). A summary of these results is given in Table 8.3. The enzyme U_2 RNase specifically cleaves phosphodiester bonds after purines. Because oligonucleotides from a complete T_1 RNase digest contain only terminal G, U_2 RNase functions as if it were an A specific nuclease. Furthermore, the phosphodiester bond between adjacent adenylic acid residues is more resistant to U_2

RNase than those between adenylic acid and any other nucleotide. As a result, in a U_2 RNase digest, oligonucleotides may end in AA as well as A when the original sequence contains consecutive A's. In Table 8.3, the U_2 RNase products also included A, UUA, and $(U_2C_2)A$, representing further digestion products of AA, UUAA, and $(U_2C_2)AA$, respectively.

The sequence of fragment A may be reconstructed uniquely from the data in Table 8.3 as follows.

1. U-5 is the right end because it contains the terminal G and it can only overlap with C-4 and C-2 and in the way shown:

```
U list                  5
        - - - -A-A-U-U-U-U-C-G-
C list              4       2
```

2. U-1 must be the left terminus because there are no bases to the left of it, and this yields

```
U list   1                 5
        A-A- - - -A-A-U-U-U-U-C-G-
C list                   4       2
```

3. All AA sequences are followed by a pyrimidine (if three consecutive A's occurred, AAA would be found in the U_2 RNase list). Therefore the only oligomer from the CMCT list that could overlap with U-1 is C-5. This also serves to position U-3 as shown.

```
U list   1    3                      5
        A-A-U-U-A-A-C- - - -A-A-U-U-U-U-C-G-
C list        5                   4       2
```

4. The remaining oligomers of the U_2 RNase list are U-2 and U-4. U-4 overlaps with C-4 at the right end of the molecule to produce

```
U list   1      3             4              5
        A-A-U-U-A-A-C- - - -(U-U-C-C) A-A-U-U-U-U-C-G-
C list          5                            4       2
```

5. The remaining U-2 can be fitted to the left of U-4 and the correct order of bases within the parentheses of U-4 is established from the

necessary overlaps with C-3 and C-1. This yields the sequence

U list 1 3 2 4 5

A-A-U-U-A-A-C-U-A-U-U-C-C-A-A-U-U-U-U-C-G-

C list 5 3 1 4 2

Next, attempts were made to obtain larger fragments that would include oligonucleotide A from the partial T_1 RNase hydrolysis of R17 RNA. As discussed above, the likelihood was very good that R17 RNA exists in solution with a more or less specific secondary structure and, as a result, only a fraction of guanylic acid residues would be susceptible to T_1 RNase hydrolysis.

The conditions of low temperature and high ionic strength were chosen for the limited digestion, and the products were separated by electrophoresis on polyacrylamide gels. As many as 40 discrete bands could be reproducibly identified, and it was estimated that the sizes of the oligonucleotides in each band ranged from mononucleotides near the bottom to fragments containing as many as 200 nucleotides near the top.

The purity of the fragments isolated from the gels was tested by treatment with T_1 RNase followed by a fingerprint of the products. In this manner, oligonucleotide A was found in bands 19 and 21 from the polyacrylamide gels. Band 21 was shown to contain two other components besides the fragment that included the oligonucleotide A sequence. It was estimated that the latter component contained between 50 and 60 nucleotides and consequently it was digested partially with T_1 RNase to produce smaller fragments. Each of these in turn was hydrolyzed completely with T_1 and pancreatic RNases. A summary of the results is shown in Table 8.4. The complete sequence can be deduced as before from the partial digestion products. It provides the codons for the amino acids in positions 81 to 99 of the coat protein.

As if in confirmation of the structural considerations that led to the isolation of the long fragment, the sequence may be written in a form showing considerable base pairing. This is shown in Fig. 8.5. Of the 24 paired bases in the structure, 19 are complementary. Most of the G residues are shown as paired with C residues which would make them very resistant to the action of T_1 RNase. Adams *et al.* (1969) have suggested that other such highly ordered base-paired structures exist

Table 8.4. Sequence analysis of large fragment containing

Oligonucleotide	T_1 RNase	Pancreatic RNase
T_{II} 17	UUCG CG	GU G 2C U
T_I 3	CAUG UUCG CG G	GGC GU AU G 2C U
T_{II} 11	CUACG AACUCCG	GAAC G AC 2U 3C
T_{II} 3	AAUUAACUAUUCCAAUUUUCG	GAAC 2AAU AC GC AAC G AC U C
T_I 6	AAUUAACUAUUCCAAUUUUCG UACUUAAAUAUG UUCG CUACG CAUG CG 2G	GGAAU AAAU GGC 2GU AAU 3AU GC 2AC U C

in R17 RNA, as can be judged by the comparatively small number of fragments obtained from partial hydrolysis with T_1 RNase.

It is also interesting to note that several of the alternative forms of some codons are used. For example, alanine is coded by (G)CA and GCU; serine, by UCG and UCC; threonine, by ACU and ACG; and

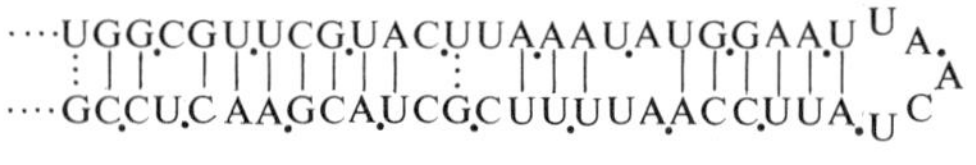

Fig. 8.5. Base-paired structure of RNA fragment from R17 coat protein cistron.

oligonucleotide A

Sequence
CGUUCG
CAUGGCGUUCG
CUACGAACUCCG
AUUAACUAUUCCAAUUUUCGCUACGAACUCCG
CAUGGCGUUCGUACUUAAAUAUGGAAUUAACUAUUCCAAUUUUCGCUACG

asparagine by AAU and AAC. This suggests that the sequences of bases in messenger RNA may not only specify the sequence of amino acids in proteins, but may also favor the formation of specific secondary structures. These structures may be required for the efficient packing of the nucleic acid into biological organelles, and perhaps also play a role in signaling the site for chain initiation.

The variability in the sequence of bases afforded by the degeneracy of the code could permit the evolutionary choice of the codons most favorable for required secondary structures. In the looped structure suggested for the fragment of the coat protein cistron, it can be seen that the third positions (the variable positions) of the codons are not opposite one another. An alteration in the third position will often yield a codon that will code for the same amino acid as before. Thus it appears that for any sequence of amino acids in proteins, a corresponding sequence of nucleotides can be written which is favorable to either an open chain or a looped secondary structure.

HYPOTHETICAL "LUCKY" OLIGONUCLEOTIDES FROM PANCREATIC RNase DIGEST

The work described above was anticipated by Kaesberg (1967), who showed the feasibility of isolating a unique 7-mer from a pancreatic digest of M12 RNA, and relating this to a portion of the coat protein sequence. In the case of MS2 RNA, which apparently is very similar to R17 RNA, Min-Jou and Fiers (1969) have isolated and determined the sequence of eight of the longest oligonucleotides from a pancreatic RNase hydrolysis. These oligonucleotides are

1. G-A-A-A-G-A-G-C-	5. A-G-A-G-A-A-G-G-U-
2. A-A-A-G-A-A-G-U-	6. G-G-G-A-A-A-A-G-G-U-
3. G-G-A-G-A-G-A-C-	7. A-A-A-G-A-G-A-G-G-A-C-
4. A-G-G-A-G-A-A-C-	8. G-A-G-G-A-G-A-A-A-G-C-

A comparison may be made between the oligonucleotides actually isolated and the hypothetical ones expected from "lucky" base substitutions in the coat protein cistron. The hypothetical oligonucleotides of chainlength eight or more are

1. GGN GGN AC
2. GG AGR AGY
3. N GGN GGN GU

It can be seen that none of the long oligonucleotides isolated from a pancreatic digest of MS-2 RNA can be part of the coat protein cistron.

APPLICATION TO OTHER RNAs

In order to apply the above procedure to other high molecular weight RNAs, the requirements are for a known amino acid sequence

Table 8.5. Hypothetical oligonucleotide from a T_1 RNase hydrolysis of TMV RNA

1.	UCN	UAY	UCN	AU3	ACN	ACN	CCN	UCN	CAR	UUY	G	31
	AGY		AGY									
2.	UN	UUY	UUR	UCN	UCN	G						15
			CUN	AGY	AGY							
3.	AR	UUR	AU3	AAY	UUR	UG						16
		CUN										
4.	N	AAY	CAR	UUY	CAR	ACN	CAR	CAR	G			23
5.	N	CAR	UUY	UCN	CAR	G						14
				AGY								
6.	AAR	CCN	UCN	CCN	CAR	G						16
			AGY									
7.	CN	AAY	CCN	ACN	ACN	G						15
8.	CN	AU3	AAY	AAY	UUR	AU3	G					18
					CUN							
9.	AR	UCN	UCN	UCN	G							12
		AGY	AGY	AGY								

Table 8.6. Hypothetical oligonucleotides from a pancreatic RNase hydrolysis of TMV RNA

1.	R	GGN	AAY		7
2.	N	AGR	AGR	GU	9
3.	3	AGR	GGN	AC	9

and a homogeneous preparation of the corresponding mRNA. Our choices are limited at present to other viral RNA and hemoglobin mRNA. When the same approach is used with the TMV RNA-TMV coat protein system, the "lucky" substitution of bases in the variable positions predicts that the oligonucleotides shown in Tables 8.5 and 8.6 may be found.

Separation of a T_1 RNase digest of TMV RNA according to chain length reveals that only three oligonucleotides with chain lengths in excess of 25 are present (Mandeles, 1968; Lloyd and Mandeles, 1970): the Ω-mer, containing 70 bases, and the two Ψ-mers containing 26 bases each (however, see Mundry, 1969). The longest hypothetical nucleotide in Table 8.5 contains 31 bases and therefore it cannot be associated with the Ω-mer. Although the two Ψ-mers are short enough to be part of the longest "lucky" oligonucleotide, there is no way of shortening this sequence to accommodate the exact chain length of the Ψ-mers. Examination of the chain length separation chromatograph pattern (Fig. 2.5) shows that there is a substantial amount of oligonucleotide material in the region above a chain length of 12 and below the Ψ-mers. It is possible that some of the sequences predicted (or portions of them) will be found in these fractions.

As for the "lucky" oligonucleotides from a pancreatic RNase hydrolysis shown in Table 8.6, the separation of oligonucleotides according to chain length from such a digest indicates that the 9-mer content is approximately 1% of the total. This indicates that as many as seven different 9-mers may be isolated from a pancreatic RNase digest of TMV RNA. The resolution of a 9-mer chain length fraction into seven individual 9-mer sequences appears to be a relatively simple experimental task (Kaesberg, 1967, Min-Jou and Fiers, 1969). A base composition analysis and a T_1 RNase digest of any one of the isolated 9-mers may serve to match it with one of the hypothetical 9-mers. For example, the hypothetical oligonucleotide N AGR AGR GU may actually occur as any one of the following:

1. A AGA AGA GU	5. G AGG AGA GU
2. G AGA AGA GU	6. G AGA AGG GU
3. A AGG AGA GU	7. A AGG AGG GU
4. A AGA AGG GU	8. G AGG AGG GU

The first, fourth, fifth, and eighth oligonucleotides are easily distinguished from the products of an alkali and T_1 RNase hydrolysis. Additional information would be needed to choose between the second and third and between the sixth and seventh possibilities. This additional information may be obtained from partial SPDE digestion and ionophoresis on DEAE-cellulose at low pH (Sanger *et al.*, 1965) or through use of a digestion with U_2 RNase (Adams *et al.*, 1969). It can be seen that the search for specific oligonucleotides corresponding to specific portions of the coat protein sequence need not present insurmountable obstacles.

SPECIFIC OLIGONUCLEOTIDES FROM HEMOGLOBIN MESSENGER RNA

Studies with rabbit reticulocytes has indicated that fairly homogeneous preparations of ^{32}P-labeled hemoglobin messenger RNA

Table 8.7. Hypothetical oligonucleotides from T_1 RNase hydrolysis of α-hemoglobin mRNA[a]

1.	UN	UUR	UCN	CCN	G									12
2.	AY	AAR	ACN	AAY	AU3	AAR	ACN	G						21
3.	N	UUY	CCN	ACN	ACN	AAR	ACN	UAY	UUY	CCN	CAY	UUY	G	35
4.	AY	UUY	ACN	CAY	G									12
5.	AR	CAR	AU3	AAR	G									12
6.	CN	UUR	ACN	AAR	G									12
		CUN												
7.	CN	UUR	UCN	ACN	UUR	UCN	G							18
		CUN			CUN									
8.	CN	CAY	AAR	UUR	CG									13
				CUN										
9.	UN	AAY	UUY	AAR	UUR	UUR	UCN	CAY	UG					25
10.	AR	UUY	ACN	CCN	G									12
11.	AY	AAR	UUY	UUR	G									12
				CUN										
12.	UN	UUR	ACN	UCN	AAR	UAY	CG							19
		CUN					AG							

[a] Chain lengths of 12 or more.

may be prepared (Chantrenne *et al.*, 1967; Labrie, 1969; Laycock and Hunt, 1969). Considering the amino acid sequence of the α and β chains of rabbit hemoglobin, the hypothetical oligonucleotides shown in Tables 8.7 through 8.10 may be produced from "lucky" substitutions of bases in the variable positions.

In the hypothetical T_1 RNase list of α-hemoglobin mRNA (Table 8.7), the third oligonucleotide contains 35 bases. There are seven positions, six N's and one R, where G may be substituted and thus shorten the length of this oligonucleotide. One or two "unlucky" substitutions may be made and still produce four oligonucleotides that contain 27 or more bases:

(a)	UUY CCN ACN ACN AAR ACN UAY UUY CCN CAY UUY G	35
(b)	ACN ACN AAR ACN UAY UUY CCN CAY UUY G	28
(c)	N UUY CCN ACN ACN AAR ACN UAY UUY CCG	28
(d)	UUY CCN ACN ACN AAR ACN UAY UUY CCG	27

The next shorter hypothetical oligonucleotide in α-hemoglobin mRNA contains 25 bases. In β-hemoglobin mRNA, the longest

Table 8.8. Hypothetical oligonucleotides from a T_1 RNase hydrolysis of β-hemoglobin mRNA[a]

1.	UN	CAY	UUR CUN	UCN	UCN	G		15
2.	AY	UUR CUN	UCN	UCN	G			12
3.	AAY	AAY	CCN	AAR	G			13
4.	N	UUR CUN	UCN	CAY	UUR CUN	G		14
5.	AY	AAY	UUR CUN	AAR	G			12
6.	CN	AAR	UUR CUN	UCN	G			12
7.	AY	AAR	UUR CUN	ACY	G			12
8.	UN	UUR CUN	UCN	CAY	CAY	UUY	G	18
9.	AR	UUY	ACN	CCN	CAR	G		15
10.	CN	UAY	CAR	AAR	G			12
11.	CN	CAY	AAR	UAY	CAY			14

[a] Chain lengths of 12 or more.

Table 8.9. Hypothetical oligonucleotides from a pancreatic RNase hydrolysis of α-hemoglobin mRNA[a]

1.	GG GAR AAR AU	10
2.	GGN GGN GAR U	10
3.	GGN AAR AAR GU	11

[a] Chain lengths of 9 or more.

Table 8.10. Hypothetical oligonucleotides from a pancreatic RNase hydrolysis of β-hemoglobin mRNA[a]

1.	N GAR GAR AAR U	11	4.	N GGN GGN GAR GC	12
	AGY	13	5.	GGN AAR AAR GU	11
2.	GG GGN AAR GU	10	6.	N AAR GGN AC	9
3.	N GAR GAR GU	9	7.	GGN AAR GAR U	10

[a] Chain lengths of 9 or more.

hypothetical oligonucleotide contains 18 bases. This means that any oligonucleotide containing 27 or more bases that can be isolated from a T_1 RNase digest of hemoglobin mRNA (presumably a mixture of the α and β forms) could be associated with a specific amino acid sequence in α hemoglobin: amino acids 36 to 46. The determination of the sequence of such an oligonucleotide could be the first demonstration of the colinearity of amino acid sequences in a protein and the corresponding base sequences in a natural mRNA.

In the hypothetical list of oligonucleotides from a T_1 RNase digest of β-hemoglobin mRNA (Table 8.8), oligonucleotide 11 contains 14 bases with one N and one R. It may therefore be shortened to yield the following oligonucleotides:

(a)	CAY AAR UAY CAY	12
(b)	UAY CAY	6

It may be noted that oligonucleotide 11, as well as the shortened sequences, terminates in a pyrimidine. Provided that β-hemoglobin mRNA ends at the codon corresponding to the last amino acid, one could expect to find this terminal sequence in the chain length fractions containing the 6-, 11- or 14-mers, if the terminal Y is phosphorylated. If the Y is not phosphorylated (or if the phosphate is removed prior to T_1 RNase action), the terminal sequence can be found in the fractions containing either 5, 11, or 13 negative charges. The location of the terminal sequence may be accomplished by the procedure of either Tomlinson and Tener (1963) or Dahlberg (1968).

According to the lists of hypothetical oligonucleotides, any of a total of 23 oligonucleotides containing 12 or more bases may be produced from a T_1 RNase hydrolysis of hemoglobin mRNA. If one such oligonucleotide does exist, what are the chances of isolating it from the digestion mixture? The sedimentation constants that have been reported for hemoglobin mRNA range from 8S to 10S (Laycock and Hunt, 1969; Chantrenne *et al.*, 1967; Labrie, 1969). Calculations of molecular weight based on these sedimentation values (Kurland, 1960; Studier, 1965) indicate that there are about 500 bases in hemoglobin mRNAs. The number of bases in tRNA and 5S RNA ranges from 75 to 120. No oligonucleotides longer than 9-mers have been found in T_1 RNase digests of these. On a statistical basis (Cantor, 1965), one

12-mer could be expected from a random RNA of the same molecular size and composition as hemoglobin mRNA. A single 12-mer represents about 2.5 to 3% of the total material in hemoglobin mRNA. This amount of oligomer is easily discernible in a chain length separation chromatogram or a two-dimensional fingerprint.

SEQUENCE ANALYSIS OF THE PROTEIN SYNTHESIS INITIATION SITES OF R17 RNA

A remarkable example of the application of the most recent technical advances as well as fundamental biochemical knowledge to the sequence analysis of the protein synthesis initiation sites of R17 RNA has been provided by Steitz (1969a,b). The techniques used were the basic fingerprint method of Sanger *et al.* (1965), as modified by Brownlee *et al.* (1968), and homochromatography essentially as described by Brownlee and Sanger (1969) (see Chapter 11 for descriptions). The sequence strategy required the isolation of specific subunits and therefore made use of an early observation of Takanami *et al.* (1965) that *E. coli* ribosomes could be used to shield specific portions of f2 RNA from nuclease attack. The plan was to bind ^{32}P-R17 RNA to ribosomes under conditions of polypeptide chain initiation and thus shield the initiation sites from nuclease digestion. After hydrolysis of the exposed nucleic acid with either T_1 or pancreatic RNase, the shielded portions could be recovered for sequence analysis.

The binding of ^{32}P-R17 RNA was carried out under conditions specified by Kondo *et al.* (1968), who found that the binding of f2 RNA to 70S ribosomes from *E. coli* in the presence of 0.005 M Mg^{2+} was strictly dependent on the presence of fMet-tRNA. With ^{32}P-R17 RNA, Steitz observed that as much as 10 to 20% of the nucleic acid was associated with the *E. coli* ribosomes under these conditions. The next step was to isolate the initiation sites of the RNA. This was accomplished by digestion of the exposed nucleic acid by either T_1 or pancreatic RNase and fractionation of the mixture by sucrose gradient centrifugation. The ribosomes from the 70S region were isolated and treated with 8 M urea in order to unfold the ribosome and help dissociate the R17 RNA fragments from the initiation complex. This mixture was then passed through a small DEAE-cellulose column to remove the bulk of the unlabeled ribosomal material, and the R17 RNA fragments were collected for sequence analysis.

In some experiments, the total fragments were digested with T_1 and pancreatic RNase and the major oligonucleotides were determined by the standard fingerprint method. Although the relative yields of oligonucleotides varied somewhat, they were generally consistent from one binding experiment to another, and they were generally the same oligonucleotides whether T_1 or pancreatic RNase was used to digest the unshielded nucleic acid. Many of these oligonucleotides could be arranged in overlapping sequences with the aid of information from the products of a partial T_1 RNase hydrolysis of the total fragments. As a result, it appeared that the primary oligonucleotides could be assigned to several discrete long sequences. One of these was implicated as the starting sequence for the coat protein cistron because it contained the sequence

A-U-G-G-C-U-U-C-U-A-A-C-U-U-U-
fMet Ala Ser Asn Phe

which codes for the N-terminal end of the R17 coat protein. This sequence was extended to include the whole of the coat protein initiation site after separation of the total fragments by homochromatography (Brownlee and Sanger, 1969). The patterns obtained in this way revealed three groups of radioactive spots, one of which was especially abundant judging from the intensity of its two components. These two major spots were isolated and their sequences were determined. It was found that they were simply size variants of the same sequence that contained the coat protein initiation site: the first spot was the smaller oligonucleotide beginning with UAA and the second spot contained a longer oligonucleotide beginning with ACG. The data were correlated into the sequence

YAG AGC CUA ACC GGG GUU UGA AGC AUG-
GGU UCU AAC UUU

The presence of two other groups of spots suggested that these may be associated with the two remaining proteins known to be coded by R17 RNA: the synthetase and the A-protein.

At this point, use was made of the observation of Lodish and Robertson (1969) that *B. stearothermophilus* ribosomes recognize and translate only the A-protein initiation site of f2 RNA. Steitz found that, when *B. stearothermophilus* ribosomes were used with R17 RNA in the initiation complex, the number of A-protein initiation fragments per mole of R17 RNA was increased tenfold. This eased considerably the sequence analysis of this fragment that was otherwise produced in very small amounts. The sequence of the A-proten initiation site was determined to be

UCC UAG GAG GUU UGA CCU AUG CGA CCU-
UUU AGU G

By elimination, the third group of spots was associated with the synthetase protein initiation site. In the sequence analysis of this group, use was made of the observation that with aged R17 RNA, the intensity of the spots corresponding to the coat protein initiation site decreased while those associated with the other two proteins increased. This was believed to be duc to the degradation of the coat protein initiation site with the concommitant greater accessibility of the synthetase and A-protein sites. As a result, more synthetase initiation fragments were produced and these could be analyzed to provide the sequence

YAA ACA UGA GGA UUA CCC AUG UCG AAG-
ACA ACA AAG

It may be noted that the amino acid sequence specified is (starting with AUG initiating codon on the left)

A-U-G-U-C-G-A-A-G-A-C-A-A-C-A-A-A-G-
fMet Ser Lys Thr Thr Lys

Recent evidence has indicated that the N-terminal tripeptide of R17 synthetase is fMet-Ser-Lys (M. Osborn, K. Weber, and H. F. Lodish, unpublished observations). The nucleic acid sequence of the initiation site predicts that, when the next tripeptide is determined it will contain

Thr-Thr-Lys

Similarly, starting with the initiating codon AUG, the amino acid sequence indicated for the N-terminal of A-protein is

A-U-G-C-G-A-G-C-U-U-U-U-A-G-U-G-
fMet Arg Ala Phe Ser

To date, fMet-Arg- has been identified as the N-terminal of R17 synthetase (Lodish and Robertson, 1969). The nucleic acid sequence at this site predicts that the beginning dipeptide will be followed by Ala-Phe-Ser and either Ala, Gly, Val, or Asp.

ADDITIONAL STUDIES ON R17 RNA

Shortly after Steitz (1969a) reported the determination of sequence of the polypeptide initiation sites of R17 RNA, two other developments were reported: one by Nichols (1970), and the other by Jeppesen *et al.* (1970). Using a combination of gel electrophoresis, thin-layer homochromatography, and the fingerprint method, Nichols established the sequence of the nucleic acid corresponding to the last six amino acids (carboxyl terminal) or the coat protein and extending to the right of that for 26 nucleotides. This portion of R17 RNA was isolated from a partial T_1 RNase hydrolysis mixture through use of gel electrophoresis and homochromatography and was found to contain 43 nucleotides. It was further degraded by T_1 RNase and the products separated by homochromatography to yield three long oligonucleotides. Each of these was then analyzed by a combination of pancreatic RNase action, CMCT-blocked pancreatic RNase digestion, and spleen acid RNase digestion. With one of the oligonucleotides, U_2 RNase aided in determining the sequence.

As a result, the following sequence was determined for this fragment:

(G)CA AAC UCC GGU AUC UAC UAA UAG AUG-
Ala Asn Ser Gly Ile Tyr
CCG CGG AUU <u>CAA ACA UG</u>

It may be noted that the underlined sequence at the right end overlaps with the left end of the fragment isolated from the synthetase initiation site by Steitz (1969a). The combined sequence for the termination of the protein coat, the intercistronic portion, and the synthetase initiation site is

(G)CA AAC UCC GGU AUC UAC UAA UAG AUG-
Ala Asn Ser Gly Ile Tyr

ACG GCC AUU CAA ACA UGA GGA UUA CCC-

AUG UCG AAG ACA ACA AAG
fMet Ser Lys Thr Thr Lys

From the results of Adams *et al.* (1969), Steitz (1969), and Nichols (1970), several things are clear:

(a) The sequence AUG is necessary although not sufficient to signal the initiation of synthesis of all three R17 RNA proteins.

(b) In a multicistronic message, termination is signaled by one or more termination codons—UAA, UAG, UGA.

(c) Termination is not directly coupled to initiation by adjacent codons, as demonstrated by the absence of termination triplets next to the initation AUG triplets for each of the R17 cistrons.

(d) An intercistronic sequence of approximately 30 nucleotides exists that may not be translated and functions in some way, as yet unknown.

(e) In R17, the synthetase cistron is to the right of the coat protein cistron.

With respect to the last result, since there are only three cistrons in R17, the remaining cistron for the A-protein must be either to the right of the synthetase or to the left of the coat protein. An approach to the resolution of this matter was provided by the isolation of a nuclease from *E. coli* MRE600, RNase IV, by Spahr and Gesteland (1968). It was found that this nuclease cleaves R17 RNA at a position about 40% of the way into the molecule from the left end to yield two fragments. *In vitro* studies with extracts from *E. coli* showed that the 60% fragment directs efficient synthesis of the synthetase protein while the 40% piece was relatively inactive as messenger. The coat protein, which is the major product of the *in vitro* protein synthesis, was not efficiently produced by either fragment. Analysis of the left ends of each fragment revealed that the 40% piece contained the left terminal pppG- in an amount that was equivalent to the undegraded R17 RNA while the

60% piece contained less than 10% pppG- terminal ends (Gesteland and Spahr, 1969). This established that the 60% piece containing the synthetase cistron was the right end and the 40% piece was the left end of R17 RNA. Because the coat protein and synthetase cistrons were contiguous and, combined, accounted for about 1800 nucleotides of the 2100 nucleotides in the 60% piece, this suggested that the A-protein cistron was in the 40% piece and to the left of the coat protein cistron (Gesteland and Spahr, 1969).

These conclusions were confirmed and the correct position of the A-protein cistron established by Jeppesen *et al.* (1970). The methods used here were essentially the same as those used by Adams *et al.* (1969) and Steitz (1969a,b) and were applied to the 40% and 60% fragments. The A-protein initiation site could be detected in the 40% fragment only, and this was especially evident when *B. stearothermophilus* ribosomes were used. The position of the coat protein cistron between the A-protein cistron and the synthetase cistron was confirmed by allocation of specific, long oligonucleotides from a T_1 RNase digest.

As a result of these studies, a biochemical map of R17 mRNA could be drawn as shown in Fig. 8.6, which summarized the sequence data.

SEQUENCE ANALYSIS OF THE COAT PROTEIN INITIATION SITE OF Qβ RNA

At the same time as Steitz (1969a), Hindley and Staples (1969) reported the sequence analysis of the initiation site of Q*β* coat protein. The experiments in this instance were performed in essentially the same way except that polyacrylamide gel electrophoresis was used to isolate the larger fragments. The results establish the following sequence for this portion of Q*β* RNA:

```
Y-A-A-U-U-U-G-A-U-C-A-U-G-G-C-A-A-A-A-U U-A-
                           fMet   Ala   Lys   Leu
G-A-G-A-C-N-
 Glu   Thr
```

Here, again, preliminary data indicate that the Q*β* coat protein begins with the sequence

fMet-Ala-Lys-Leu- (M. Osborn, unpublished).

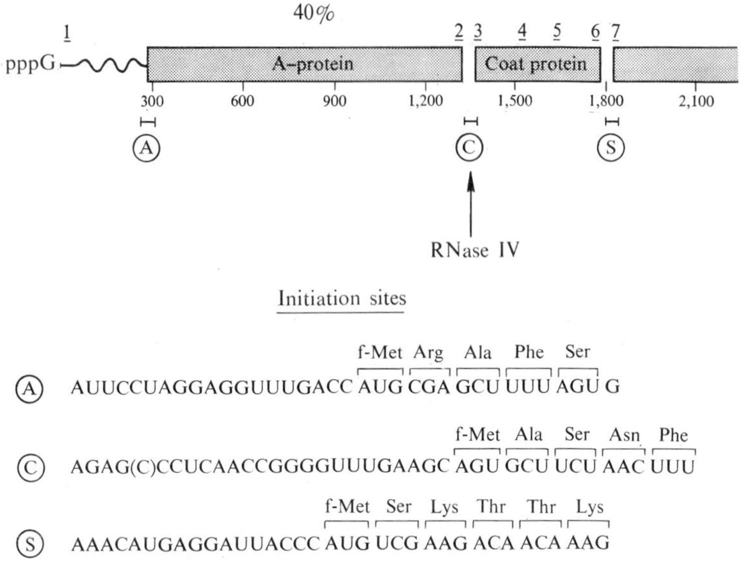

Fig. 8.6. Summary of R17 RNA sequence data (redrawn Jeppesen *et al.*, 1970).

The nucleic acid sequence predicts that this tetrapeptide will be followed by Glu-Thr.

REPLICATION METHODS

Another method for limiting the sequence analysis to a specific segment of RNA has utilized the bacteriophage replicative mechanism (Billeter *et al.*, 1969). In the case of the virus Qβ, it is known that a replicase specific for Qβ RNA is elaborated during the infection of *E. coli*. It appears that this replicase is formed as a result of the

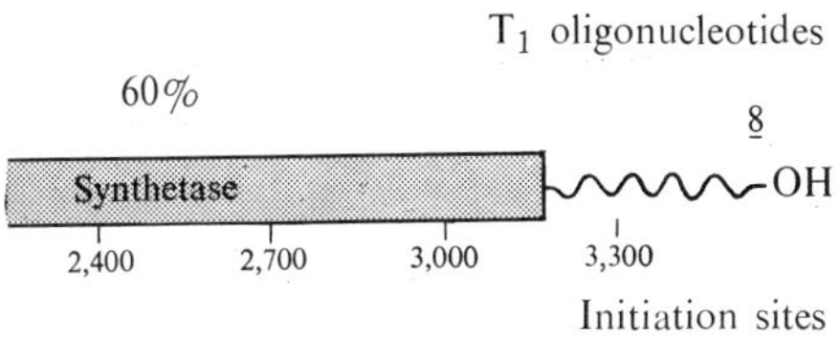

T$_1$ oligonucleotides

1 CCAUUUUUAAUG

2 (C)CCUCAACCG

Ala Ser Asn Phe Thr Gln
3 CU UCU AAC UUU ACU CAG

Lys Tyr Thr Ile Lys
4 C AAA UAC ACC AUU AAA G

Glu Leu Thr Ile Pro Ile Phe
5 AA UUA ACU AUU CCA AUU UUC G

6 CCAUUCAAACAUG

Thr Thr Lys
7 ACA ACA AAG

8 UUACCACCCA$_{OH}$

translation of a portion of the Qβ RNA into protein and that it will function as a replicase only with Qβ RNA as a template. *In vitro*, Qβ replicase catalyzes the synthesis of a single strand (minus strand) of RNA complementary to Qβ RNA (plus strand). The newly synthesized minus strands in turn serve as templates for the synthesis of progeny plus strands from which the progeny virus particles may be formed. It has been shown by Franze de Fernandez *et al.* (1968) that, if the replicase is purified sufficiently to remove certain factors normally present (Shapiro *et al.*, 1968), Qβ plus strands no longer serve as templates, while minus strands are fully active. This would permit the use of a mixed

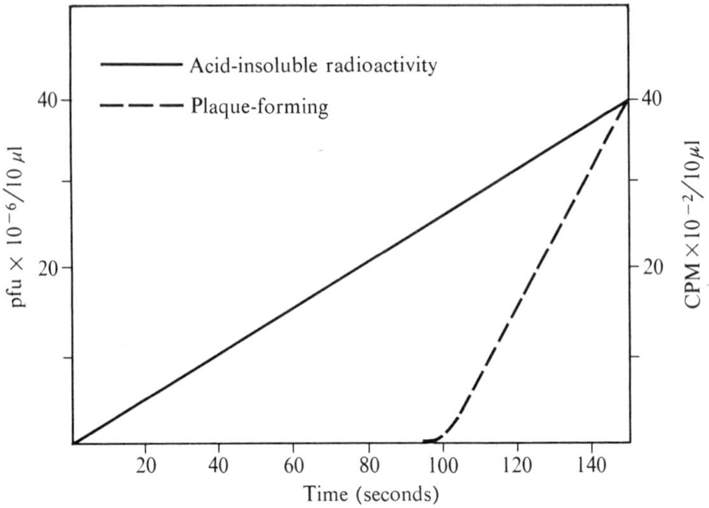

Fig. 8.7. Comparison between the formation of infectious R17 RNA and the incorporation of radioactive nucleotide.

population of plus and minus strands from the denaturation of double-strand Qβ RNA (Weissman *et al.*, 1967) as a source of templates for the synthesis of plus strands only, and provides a system for the synthesis of a specific RNA.

The next problem was how to limit the synthesis of the RNA to specific segments for sequence analysis. The information needed to devise an appropriate plan came from two sources. One was a study of the time course of production of infectious Qβ RNA when minus strands are used as templates. In this study, Qβ RNA minus strands were preincubated with Qβ replicas and ATP for 2 minutes. Synthesis was then initiated by the addition of the remaining three nucleoside triphosphates, one of which was radioactive. The increase in infectivity and incorporation of radioactivity are shown in Fig. 8.7. Within a few seconds after the addition of the remaining three nucleoside triphosphates, incorporation of the radioactivity became linear with time and remained so throughout the course of the reaction. However, no infectious RNA was produced until 100 seconds had elapsed: after that, this rate became linear with time also. Because Qβ RNA contains approximately 3300 bases, the average rate of chain propagation appears to be at least 3300/100 = 33 nucleotides per second at 37°.

Because the replication of Qβ RNA proceeds from left to right, the second piece of information needed was the sequence of bases at the left end. This was provided by DeWachter and Fiers (1969), who found that the left terminus was

pppGpGpGpGpApApCp . . .

On this basis, Billeter and his co-workers elected to preincubate enzyme and template with GTP and ATP in order to obtain initiation and polymerization up to the position where CTP is required for further synthesis. This serves to saturate all of the chain initiation sites with a short known sequence of nucleotides and provides the basis for the subsequent synchronous synthesis of the remaining RNA.

The synthesis of a labeled sequence of about 150 nucleotides would require 4 to 5 seconds at 37°. In order to improve the control of the reaction, the temperature was lowered to 20° and the nucleotide concentration was reduced from the optimum of 0.8 m*M* to 0.1 m*M*. As a result, the polymerization rate was reduced to about 5 nucleotides per second. Upon addition of the missing nucleoside triphosphates under these conditions, a linear incorporation occurred immediately with no evidence for new chain initiation within the next minute. From this evidence it was reasonable to assume that suitable specific short segments of Qβ RNA could be produced by limiting the synthesis time to fractions of a minute.

Experiments were subsequently carried out for periods under 1 minute with each of the ^{32}P-nucleoside triphosphates as well as all four simultaneously. A typical 30-second product (which should contain 150 bases if the synthetic rate is 5 per second) had a sedimentation coefficient of 5.5S, and it was estimated that this segment contained about 140 nucleotides. Base composition analysis of the same segment indicated a chain length of about 160. A segment of this size is within the capabilities of present known sequence analysis methods.

The labeled products were analyzed according to the methods of Sanger *et al.* (1965), and the use of ^{32}P-α-labeled nucleoside triphosphates in the synthesis permitted the use of nearest neighbor frequency data as well. An example of this procedure is given in Table 8.11. Here

Table 8.11. Sequence analysis of fragment T16

	Labeled Substrate	Alkaline Hydrolysis		Pancreatic RNase	Sequence Deduced
1.	ppp* U ppp* A ppp* C ppp* G	2.7Up*, 1.8Cp*,	1.0Ap* 1.0Gp*	Ap*Gp*, Up*, Cp*	(U3,C2)p*Ap*Gp*
2.	ppp* A	1.2Cp*,	1.0Gp*,	Ap*Gp*, Cp*,	(U3C)pCp*Ap*Gp*(A)
3.	ppp* U	Up*		Up*	Up*Up*UpCpCpApGp(A)
4.	ppp* C	1.0Up*,	1.0Cp*	Up*, Cp*	UpUpUp*Cp*CpApGp(A)
5.	ppp* G	Ap*		Ap*Gp	UpUpUpCpCpAp*Gp(A)

oligonucleotide T16 was isolated from a T_1 RNase hydrolysis of a 30-second segment in five separate experiments. In one experiment, all four nucleoside triphosphates were labeled in the α position with ^{32}P, and in the remaining experiments only one of the four nucleoside triphosphates was labeled. From Table 8.11, it may be seen that the fourth and fifth experiments confirm the deductions of the first three. The same analytical procedure was applied to the segments formed during shorter periods of replication. Oligonucleotides obtained from pancreatic or T_1 RNase digestion of these segments could be tabulated with two additional pieces of information: the identity of the nearest neighbor to the right of the oligonucleotide, and the time of incubation necessary to form the segment in which the oligonucleotide first appeared. These data are shown in Table 8.12.

As an example of the procedure for aligning the oligonucleotides into a longer sequence, we can consider the pancreatic RNase fragment P19:

A-G-G-A-G-C-

This oligonucleotide was found in the replication segment formed after 20 seconds of incubation. Because it is an oligonucleotide from a pancreatic RNase digestion, it is preceded by a pyrimidine and, from the nearest neighbor data, it is followed by a U:

Y-A-G-G-A-G-C-U-
19

Table 8.12A. T_1 RNase products of 30-second product

Spot No.	Moles (at 30 sec)	Standard Time of Appearance (sec)	Sequence
1a	3–6	3	G [G]
1b	3	early, late	G [A]
1c	2	late	G [U]
2	2	13, 18	CG [U]
3	1	23	CCG [A]
4	1	9	CAG [U]
5a	1–2	12	AG [G]
5b	2	23, 8 or 23	AG [C]
5c	1	8 or 23	AG [U]
7	<1	>30	ACACG [C]
8	1	17	UG [G]
9	<1	>30	(CAU)(C)CG [U]
10	1	21	UCUG [C]
11	1	24	CUCUG [G]
12	1	24	AUAAUG [A]
13	1	13	UAUAAG [A]
14	1	15	ACAUAUG [C]
15	1	9	UC(ACACACC)(UC)AG [C]
16	1	24	UUUCCAG [A]
17	1	27	AAUCUUCCG [A]
18	1	15	CCUAAAUUACCG [C]
21	1	23	UUUCG [G]
22	1	23	AUUUUCAG [G]
23	1	23	AAAUU(U)(CU)AAUG [A]
24a	1	11	UACUUCACUG [A]
24b	1	~25	ACCU$_n$(CU$_m$)(AU)CG [A]
25	1	3	ACCCCCCUUUAG [G]
26	1	2	pppGp [G]

The T_1 RNase oligonucleotides that obviously overlap are T_{1b} and T_{5b}, each of which appears after 20 seconds of replication:

$$\overset{}{\text{Y-A-}}\underbrace{\overline{\text{G}}^{\,1}\text{-}\overline{\text{A-G}}^{\,5}\text{-C-U-}}_{19}$$

The oligonucleotide required for the correct overlap on the left of P19 must have the terminal sequence

. . . Y-A-G(G)

Table 8.12B. Pancreatic RNase products of 30-second product

Spot No.	Moles	Standard Time of Appearance (sec)	Sequence
1a	1–2	5–10	U [G]
1b			U [C]
1c			U [A]
1d			U [U]
2a	1	10–15	C [G]
2b			C [C]
2c			C [A]
2d			C [U]
4a	1	28	AC [G]
4b			AC [C]
4c			AC [A]
4d	2		AC [U]
5a	2–3	13, 22	GC [G]
5b	1		GC [C]
6a	1	13	AU [G]
6b	1		AU [C]
6e	2		AU [A]
7	1	5	AGC [A]
8	2	21, 28	AAU [G]
9a	2	15, 30	GU [G]
9b	1		GU [U]
10	<1	26	AGAC [C]
11	1		AAAU [U]
12a	1		GAU [A]
12b	1		GAU [U]
13	1	5–10	AGU [A]
14a	1–2	15–20 (14a–14b)	GGU [C]
14b	1		GGU [U]
15	1	8	GAGU [A]
16	<1		GAAU [C]
17	1		GAAAU [U]
18	1	23	GGAGC [C]
19	<1	24	AGGAGC [U]
20	1	12	(AAG)(AGG)AC [A]
21	1	<5	pppGGGG$_{1-2}$AC [C]
22	1	~5	AGGGGGU [?]

Both T22 and T25 have this terminal sequence. T25 must be eliminated because it is found in a segment formed after only 3 seconds of replication and it therefore cannot be near P19. T22 is found in a segment after 20 seconds of replication and therefore provides the correct overlapping sequence for the left end:

22 1 5 11 1

(G)A-U-U-U-U-C-A-G-G-A-G-C-U-C-U-G-G-

19

In this manner, the sequence of a segment containing 175 nucleotides was reconstructed as shown in Fig. 8.8. From this sequence it can be concluded that the protein coat cistron of Qβ does not begin in the segment analyzed. Billeter and co-workers point out that it should be technically possible to continue the sequence analysis of Qβ RNA in a similar fashion by synthesizing a segment with unlabeled nucleoside triphosphates for the first 30 seconds, followed by synthesis with labeled nucleoside triphosphates for the next 30 seconds. The limiting factor would be the maintenance of synchrony for the longer periods of synthesis.

GENERALIZATION OF REPLICATIVE METHODS

In order for the same kind of sequence analysis to be carried out on other viral RNAs, the analogous specific RNA replicases must be known. At present, Qβ replicase is the only specific replicase that has been purified and characterized enough for this purpose.

It is possible, however, to use the relatively nonspecific DNA template-requiring RNA polymerases for the formation or specific segments. The main requirements for such a system would be for a DNA template with a single initiation site and RNA polymerase preparation with a preference for initiating synthesis of RNA at that site. These criteria would seem to be met in the case of ϕX 174 or Fd DNA and *E. coli* polymerase (containing σ factor). In the case of Fd DNA, Okamoto *et al.* (1970) have found that there are three initiation sites, only one of which produces a strand of RNA that is long enough to be a copy of the viral DNA. The other two sites produce RNA strands that are roughly two-thirds and one-third of the size of

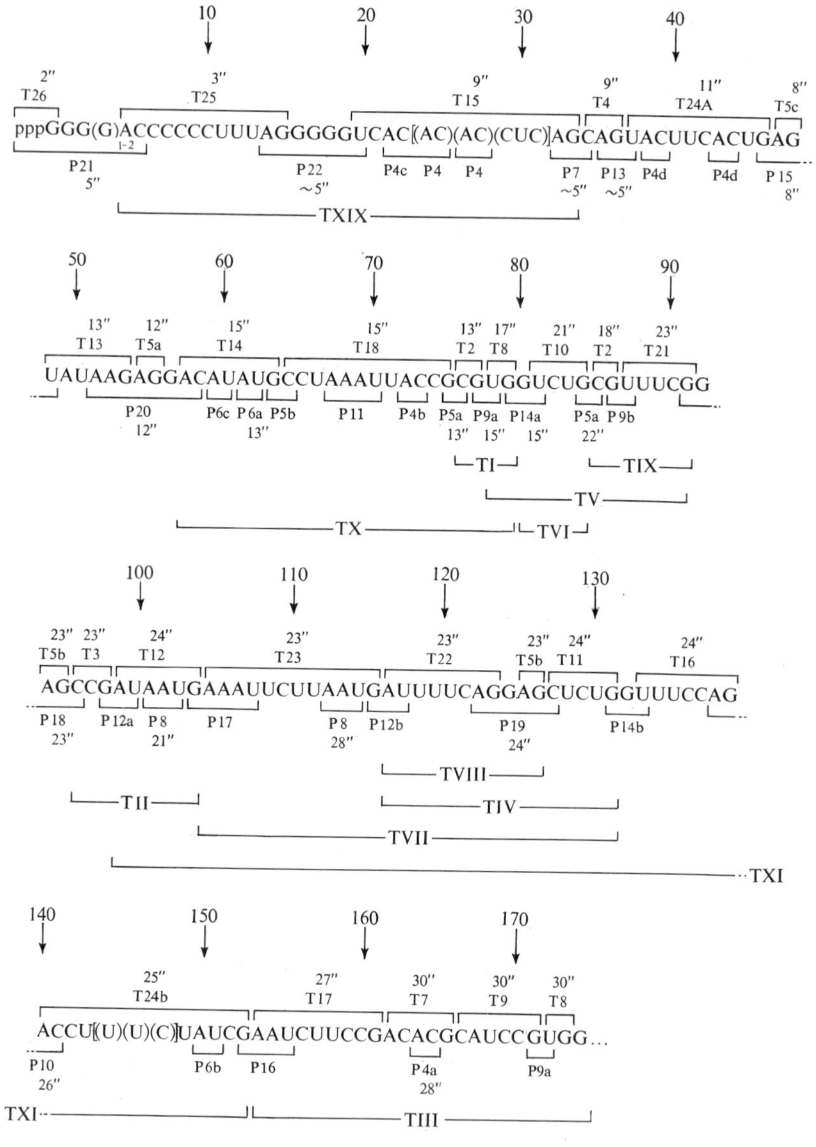

Fig. 8.8. Sequence of "30-second" product of *in vitro* replication of Qβ RNA. Shown also are the portions obtained from shorter incubation times, ———, which aided materially the sequence analysis (from Billeter *et al.*, 1969).

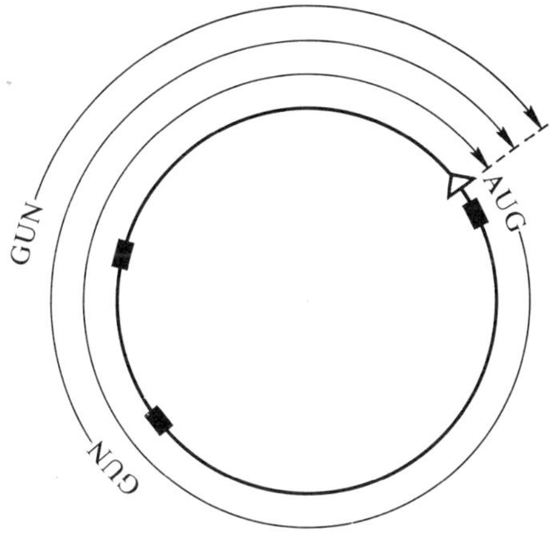

Fig. 8.9. Model of transcription on RF-I DNA from coliphage Fd; N represents either A or U (from Suguira *et al.*, 1969).

the first. The sequence of the bases at the main initiation site was determined to be AUG, and that at the two-thirds and one-third positions GUA and GUU, respectively (Sugiura *et al.*, 1969). This is shown diagrammatically in Fig. 8.9. Initiation at the main site was preferred when the polymerase preparation contained the σ factor and that initiation would occur at the other sites equally well when the σ factor was removed from the polymerase by molecular size exclusion chromatography on Sephadex (Sugiura *et al.*, 1969). Synchrony and synthesis of short segments presumably could be achieved in a manner analogous to that performed with the Qβ replicase-Qβ RNA system.

An alternative that would not require synchrony is provided by the use of the γ-^{32}P-labeled nucleoside triphosphate corresponding to the left terminal base. If, as in the case of Fd DNA, the first three bases synthesized are pppApUpGp . . . , ^{32}pppA could be used to distinguish all segments that begin with this sequence. If the unsynchronized replication reaction could be induced to terminate at a specific base, a series of oligonucleotides would be obtained whose chain lengths would indicate the positions of the terminating base. Chain termination at specific positions could be accomplished either through use of 3′-deoxynucleoside triphosphates as chain terminators or by limiting the concentration of one of the reactant nucleoside triphosphates. In common with most other end label methods (Chapter 7), the utility of such a procedure is restricted by the ability to separate oligonucleotides of chain length N from $N + 1$.

Another approach that utilizes the replicative mechanism has been suggested by Burton (1965). It entails the use of a pulse of radioactive nucleoside triphosphate during the unsynchronized replication reaction. The position of specific, unique oligonucleotides with respect to the right end could be determined by their relative radioactivity: oligonucleotides near the right end should contain more radioactivity per base than those farther from the right end.

PARTIALLY STRIPPED TMV

The use of partially stripped TMV for sequence analysis is based on the original observation of May and Knight (1965) that, in the presence of sodium dodecyl sulfate (SDS), the coat proteins of TMV are removed in a polar fashion: the proteins are dislodged from around the inner nucleic acid core starting from the 5′-OH-linked (right) terminus of the nucleic acid and proceeding toward the 3′-OH-linked (left) end. The supporting evidence included electron photomicrographs of partially stripped TMV which showed nucleic acid protruding from one end only (Hart, 1955, 1958). If the stripping reaction could occur at either end or in the middle, the electron photomicrographs would be expected to show stripped virus with protein missing from the center or from both ends, as is found in the case of removal of protein from TMV in the presence of phenol (Corbett, 1964).

In addition, it was found that, if partially stripped TMV was treated with either SPDE or VPDE and then reconstituted with TMV protein, almost all of the infectivity was lost in the VPDE-treated samples while normal infectivity was retained in the SPDE-treated samples. This indicates that the stripping reaction is indeed polar and that it proceeds from the right end toward the left.

This phenomenon of polar stripping in the presence of SDS was used by Kado and Knight (1966) for the location of the local lesion gene in TMV. The TMV was partially stripped to different extents in separate samples and each was treated with dilute nitrous acid. Under these conditions, the oxidation of the exposed nucleic acid was limited in the main to the conversion of a few cytosine bases into uracil bases. Each sample of stripped virus was then reconstituted with TMV protein, and the chemically mutated TMV was used to infect a local lesion host, *Nicotiana sylvestris*. Unaltered TMV ordinarily produces a green mosaic pattern upon infection of this host, while nitrous acid-treated TMV often produces local lesions. The appearance of local lesions was therefore correlated with the extent of stripping, and the position of the local lesion gene was expressed as a percentage of the total distance from the stripped end. The local lesion gene was located in the left half of the molecule.

In a somewhat different manner, the location of the cistron for the coat protein has been estimated (Kado and Knight, 1968). In this instance, a common strain of TMV, *vulgare*, was partially stripped with SDS, and the exposed nucleic acid was hydrolyzed with pancreatic RNase. These partial virus particles were then mixed with a preparation of whole TMV of the HR strain, and the mixture was used to infect tobacco plants. It was found that TMV *vulgare* coat protein could be detected immunologically in plants infected with the mixture of TMV HR and 50% stripped TMV *vulgare* or with whole TMV *vulgare*, but not with TMV HR alone or the 50% stripped TMV *vulgare* alone, or with a mixture of TMV HR with TMV *vulgare* stripped more than 70%. This finding provided evidence that the protein coat gene was also located in the left half of TMV RNA.

LOCATION OF UNIQUE OLIGONUCLEOTIDES

TMV that has been partially stripped with SDS has also been used to locate unique oligonucleotides from T_1 RNase hydrolyses (Mandeles, 1968). As indicated before, one of these oligonucleotides contains 70 bases and has been designated the Ω-mer because it is eluted last in a chain length column chromatographic separation of a T_1 RNase digest. The other two contain 26 bases each, are eluted next to last in a chain length chromatogram, and are designated the Ψ-mers. One of these is eluted first in an acid column and is designated the Ψ_1-mer, the other is eluted second and is designated the Ψ_2-mer. The molar yield of each oligonucleotide indicates that each occurs only once in TMV RNA (Lloyd and Mandeles, 1970).

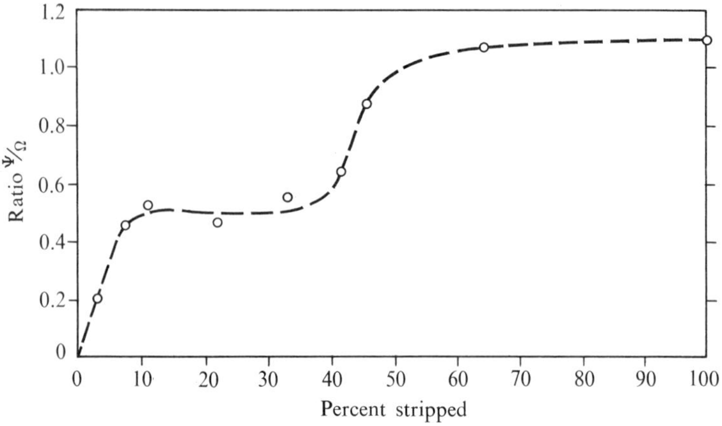

Fig. 8.10. Relation between the ratio of Ψ-mer (combined Ψ_1 and Ψ_2) to Ω-mer and the percentage of TMV RNA exposed.

The positions of each of these three unique oligonucleotides were located in the original TMV RNA molecule in the following way: whole TMV was partially stripped with SDS, the exposed nucleic acid was hydrolyzed with T_1 RNase, and the digestion mixture separated according to chain length by column chromatography. The chromatographic pattern was examined for the presence of the unique oligonucleotides, and it was found that as the extent of the stripping progressed, the Ω-mer was first exposed, then the Ψ_1-mer, and last the Ψ_2-mer. When the ratio of the amount of Ψ-mer to Ω-mer is plotted against the degree of stripping in each trail, a curve is obtained as shown in Fig. 8.10. For completely stripped TMV, the value of this ratio is 1.1. From Fig. 8.10, it appears that most of the Ω-mer was exposed after 3% stripping and that the Ψ:Ω ratio is very low. This implies that the Ω-mer is located closer to the right end of TMV RNA than are the Ψ-mers, and that the left end of the Ω-mer is probably within 180 bases of the right end of TMV RNA. As more of the TMV RNA is exposed, the ratio of Ψ-mers to Ω-mer increased until, at 12% stripped, the ratio is 0.5 that of the value obtained when stripping is complete. This indicates that one of the Ψ-mers has been completely exposed at this point and that it is probably located at about 6% or 380 bases in from the right end. This ratio remains constant to 35% stripped, after which

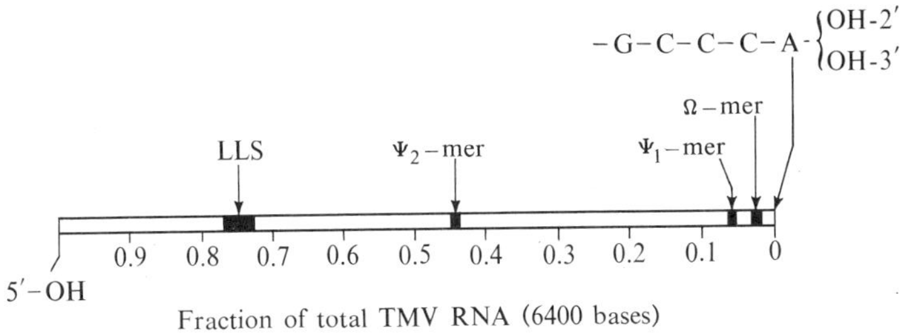

Fig. 8.11. Summary of sequence data for TMV RNA. Shown are the positions and relative sizes of Ψ_1-mer, Ψ_2-mer, and Ω-mer. LLS is the local lesion gene identified by Kado and Knight (1966).

it again rises and eventually equals the ratio of completely stripped TMV. From the graph, it may be inferred that the second Ψ-mer is located at about 44% of the distance in from the right end.

When the first Ψ-mer exposed is chromatographed at low pH it is eluted early, identifying it as Ψ_1-mer. A summary of these results is shown in Fig. 8.11. This, then, provides a method for estimating the position of unique oligonucleotides from a T_1 or pancreatic RNase digestion mixture.

APPLICATION TO SEQUENCE ANALYSIS

From the work of Kaesberg (1967) and Min-Jou and Fiers (1969), it is clear that unique sequences with chain lengths as short as 8 are found in pancreatic digests of M12 or MS2 RNA. However, the use of the stripping reaction need not be limited to the location of unique sequences. Oligonucleotides containing as few as four bases may also be located in the TMV RNA molecule. Consider, for example, the tetramers from a T_1 RNase digest of TMV RNA. This fraction comprises 8% of the total nucleotides in TMV RNA, or about 512 nucleotides. Expressed as the number of tetramers, this is 512/4 or 128 separate tetramers. In a T_1 RNase digest, only 27 different sequence isomers are possible. If equal amounts of each of the possible teramers occurred, there would be 128/27 or about five of each kind present. The partial stripping of TMV could be used to establish the approximate position of each of this comparatively small number of oligonucleotides. If, in addition, the analysis was performed with ^{32}P-labeled TMV RNA and the two-dimensional ionophoretic method described in Chapter 4, the relative positions of the tetramers in TMV RNA could be determined from measurements of the relative radioactivity. Furthermore, the relative position could be estimated for any oligonucleotide isolated from the two-dimensional fingerprint. In this regard, it would seem that accurate information about the extent of the stripping reaction is less important for the determining the relative positions of specific oligonucleotides than it is for establishing the absolute positions.

A preliminary study has indicated that the stripping reaction may be limited to a small but precisely defined portion of TMV. As little as 1 to 2% of the protein can be removed with very dilute (0.05%) SDS at low temperatures, thus exposing approximately 64 to 128 nucleotides. A sample of partially stripped virus (PSV) of this kind could be divided in half, and one portion digested with T_1 RNase while the other is digested with pancreatic RNase. The small number of oligonucleotides in each digest may be identified and compared for overlapping sequences. Portions of the stripped virus could also be held for partial digestion with either of the two nucleases to provide additional overlapping sequences in the same way that sequence analysis was performed with tRNA and 5S RNA. The extent of stripping could be varied in subsequent experiments to uncover more of the molecule for additional sequences, or less of the molecule to resolve any ambiguities that arise because of the imprecision of the stripping reaction.

For more extensive stripping, the results of Symington and Commoner (1967) and of Symington (1969) indicate that a preliminary fractionation of the PSV particles may be necessary to achieve an appropriately narrow distribution of exposed chain lengths for each sequence analysis.

REFERENCES

Adams, J. M., Jeppesen, P. G. N., Sanger, F., and Barrell, B. G. (1969). *Nature*, **223**, 1009.

Apgar, J., Everett, G. A., and Holley, R. W. (1965). *Proc. Natl. Acad. Sci.*, **53**, 546.

Bassel, B. A., and Spiegelman, S. (1967). *Proc. Natl. Acad. Sci.*, **58**, 1155.

Billeter, M. A., Dahlberg, J. E., Goodman, H. M., Hindley, J., and Weissmann, C. (1969). *Nature*, **244**, 1083.
Brownlee, G. G., Sanger, F., and Barrell, B. G. (1968). *J. Mol. Biol.*, **34**, 379.
Brownlee, G. G., and Sanger, F. (1969). *Eur. J. Biochem.*, **11**, 395.
Burton, K. (1965). In *Essays in Biochemistry*. P. N. Campbell, and F. D. Greville (eds.). Academic Press, New York.
Cantor, C. R. (1965). Ph.D. Thesis, Dept. of Chemistry, University of California, Berkeley.
Chantrenne, H., Burny, A., and Marbaix, G. (1967). *Prog. Nucl. Acid Res. Mol. Biol.*, **7**, 173.
Corbett, M. J. (1964). *Virology*, **22**, 539.
Cox, R. A. (1969). *Biochem. J.*, **114**, 743.
Cox, R. A., and Bonanou, S. A. (1969). *Biochem. J.*, **114**, 769.
Dahlberg, J. E. (1968). *Nature*, **220**, 548.
Franze de Fernandez, M. T., Eoyang, L., and August, J. T. (1968). *Nature*, **219**, 588.
Gesteland, R. F., and Spahr, P. E. (1969). *Cold Spring Harbor Symp. Quant. Biol.*, **34**, 707.
Gould, H. (1966). *Biochemistry*, **5**, 1103.
Gould, H., Pinder, J. C., Matthews, H. R., and Gordon, A. H. (1969). *Anal. Biochem.*, **29**, 1.
Hart, R. G. (1955). *Proc. Natl. Acad. Sci.*, **41**, 261.
Hart, R. G. (1958). *Biochim. Biophys. Acta*, **28**, 457.
Hindley, J., and Staples, D. H. (1969). *Nature*, **224**, 964.
Jeppesen, P. G. N., Steitz, J. A., Gesteland, R. F., and Spahr, P. F. (1970). *Nature*, **226**, 230.
Kado, C. I., and Knight, C. A. (1966). *Proc. Natl. Acad. Sci.*, **55**, 1276.
Kado, C. I., and Knight, C. A. (1968). *J. Mol. Biol.*, **36**, 15.
Kaesberg, P. (1967). In *The Molecular Biology of Viruses*. J. S. Colter and W. Paranchych (eds.). Academic Press, New York.
Kondo, M., Eggerston, G., Eisenstadt, J., and Lengyel, P. (1968). *Nature*, **220**, 368.
Kurland, C. G. (1960). *J. Mol. Biol.*, **2**, 83.
Labrie, F. (1969). *Nature*, **211**, 1217.
Laycock, D. G., and Hunt, J. A. (1969). *Nature*, **221**, 1118.
Lloyd, D. A., and Mandeles, S. (1970). *Biochemistry*, **9**, 932.
Lodish, H. F., and Robertson, H. D. (1969). *Cold Spring Harbor Symp. Quant. Biol.*, **34**, 655.
McPhie, P., Hounsell, J., and Gratzer, W. G. (1966). *Biochemistry*, **5**, 988.
Mandeles, S. (1968). *J. Biol. Chem.*, **243**, 3671.
May, D. S., and Knight, C. A. (1965). *Virology*, **25**, 502.
Min-Jou, W., and Fiers, W. (1969). *J. Mol. Biol.*, **40**, 187.
Mundry, K. W. (1969). *Mol. Genetics*, **105**, 361.
Nichols, J. L. (1970). *Nature*, **225**, 147.
Okamoto, T., Sugiura, M., and Takanami, M. (1970). *Biochemistry*, **9**, 3533.
Penswick, J. R., and Holley, R. W. (1965). *Proc. Natl. Acad. Sci.*, **53**, 543.
Sanger, F., Brownlee, G. G., and Barrell, B. G. (1965). *J. Mol. Biol.*, **13**, 373.
Shapiro, L., Franze de Fernandez, M. T., and August, J. T. (1968). *Nature*, **220**, 478.
Spahr, P. F., and Gesteland, R. F. (1968). *Proc. Natl. Acad. Sci.*, **59**, 876.
Spahr, P. F., Farber, M., and Gesteland, R. F. (1969). *Nature*, **222**, 455.

Steitz, J. A. (1969a). *Nature*, **224**, 957.
Steitz, J. A. (1969b). *Cold Spring Harbor Symp. Quant. Biol.*, **34**, 621.
Studier, F. W. (1965). *J. Mol. Biol.*, **11**, 373.
Sugiura, M., Okamoto, T., and Takanami, M. (1969). *J. Mol. Biol.*, **43**, 299.
Symington, J. (1969). *Virology*, **38**, 309.
Symington, J., and Commoner, B. (1967). *Virology*, **32**, 1.
Takanami, M., Yan, Y., and Jukes, T. H. (1965). *J. Mol. Biol.*, **12**, 761.
Thach, S., and Boedtker, H. (1969). *J. Mol. Biol.*, **45**, 451.
Tomlinson, R. V., and Tener, G. M. (1963). *Biochemistry*, **2**, 703.
Weber, K. (1967). *Biochemistry*, **6**, 3144.
Weissmann, C., Feix, G., Slor, H., and Pollet, R. (1967). *Proc. Natl. Acad. Sci.*, **57**, 1870.

9.

PHYSICAL METHODS

The ability to discern or measure some sequence-dependent physical property provides the basis for several different methods of nucleic acid sequence analysis. They include electron microscopy, optical rotatory dispersion and circular dichroism, X-ray diffraction, nuclear magnetic resonance, and mass spectrometry. In theory, the main advantages of these methods are:

1. Analyses are performed on the intact molecule.
2. Few manipulations are necessary beyond the purification of the intact molecule.
3. Small amounts of material are sufficient.
4. Generally, only a few analyses are needed.

Despite these positive features, none of the methods (with the exception of ORD) has played any significant role in a base sequence determination. This is probably due to the requirement for sophisticated and expensive instrumentation: few investigators with access to the instrumentation and who are expert in the measurement and interpretation of physical properties have committed their research programs to nucleic acid sequence analysis. As a result, the chemical and biochemical methods described in the previous chapters have produced most of the sequence information.

The potential is high, however, for physical methods such as electron microscopy to make substantial contributions to nucleic acid sequence analysis. It is hoped that the following will help to stimulate increased interest in the use of physical approaches to sequence problems.

ELECTRON MICROSCOPY

The fundamental idea in nucleic acid sequence analysis by electron microscopy is the direct visualization of the bases *in situ* on a polynucleotide chain. Three major factors influence the implementation of this plan:

1. The resolving power of the electron microscope relative to the internucleotide distances and to the sizes and shapes of the individual bases.
2. The low electron density of polynucleotide chains.
3. The presence of background structures of the same general size and shape as the various portions of nucleic acids.

Most commercially available electron microscopes can resolve structure down to the 5 Å level. This means that the instrument is capable of showing a separation of 5 Å level between two structures that are each 5 Å in width, but that such a separation is not discernible when the distance is decreased to 4 Å or less. Structures closer than 5 Å appear fused into an indistinct mass. It is theoretically possible to construct electron microscopes with greater resolving power, and instruments have been built that can distinguish structures as close as 2 Å. Whether this resolution would be sufficient to identify individual bases in a polynucleotide chain may be seen by consideration of a portion of a nucleic acid shown in Fig. 9.1. The nucleic acid is drawn in an extended configuration to provide a 7 Å or near maximum degree of separation between the bases. Purine bases measure 2.5 Å and 5.5 Å at the shortest and longest diameters, respectively. Pyrimidine bases measure 2 Å and 4 Å. The resolution provided by commercially available instruments would not fall far short of that required for sequence analysis.

The next consideration is the low electron density of the polynucleotide chain. A number of techniques have been devised to render biological structures less transparent to an electron beam. They include: shadow casting and replica techniques and positive and negative

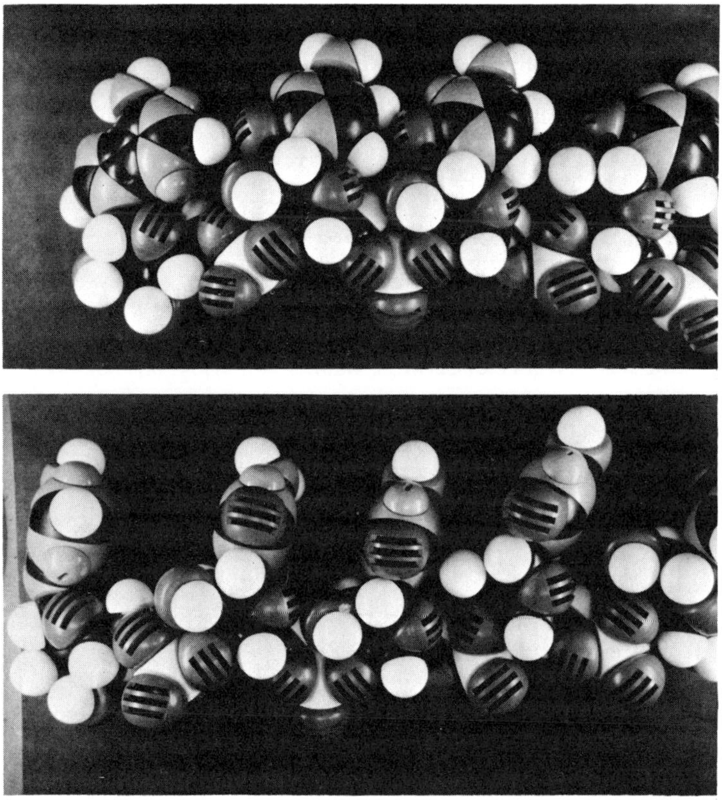

Fig. 9.1. Nucleotide spacing in nucleic acids. *Top*, plane of bases parallel to nucleic acid backbone results in a slight curvature of chain. *Bottom*, plane of bases perpendicular to nucleic acid backbone permits closer spacing between bases and straight chain. One inch equals 10 Å.

staining. Shadow casting (and replica techniques) involves the evaporation of a metal of high atomic number onto a specimen at a predetermined angle. The metals that have been used are: platinum, platinum-palladium alloy, uranium, and tungsten. Of these, tungsten is reported to provide the finest detail, of the order of 8 Å resolution. It is also more difficult to use and is less contrasty than the others. In general, the granulation produced by shadow casting is too large to permit the resolution of the individual bases but small enough to permit the determination of the size and shape of larger structures such as viruses.

The use of heavy metal stains appears to be more promising for nucleic acid sequence analysis. Early work by Zobel and Beer (1961) and Beer and Zobel (1961) demonstrated the effectiveness of uranyl acetate as a DNA stain. It was found that, in stained DNA, one uranyl ion with a diameter of 3 Å occurred at an average distance of 3.4 Å along the polynucleotide chain. Although the structural details of the various portions of the nucleic acid were totally obscured, the stained DNA was discernible against an irregular background. The results indicated that, on a noise-free background, a single atom of uranium should provide enough contrast to be used as a marker, and that such markers could be used in sequence analysis if they were base specific.

Considering that the best carbon support made from a mica surface will always show a pebbly granularity of about 5 Å (Griffith, 1969) and that uranyl acetate is nonspecific in its attachment to nucleic acid chains, a simple staining procedure is obviously inadequate for sequence analysis. In answer to these problems, Beer and Moudrianakis (1962) suggested that a bifunctional adapter reagent be used to combine with a specific base in a polynucleotide chain as well as with two or more uranium atoms for the requisite electron density. The plan then would be to make an electron photomicrograph of the nucleic acid modified in this manner, and to measure the distances between the marked bases. If this procedure could be carried out for at least three of the four bases in nucleic acids, the sequence would be determined.

One of the first reagents examined for this purpose was 8-diazonium-1,3,6-naphthalenetrisulfonic acid (Beer and Moudrianakis, 1962), shown in Fig. 9.2, which was coupled with DNA at high pH. The marked DNA was then picked up on carbon-coated grids and stained with uranyl acetate. Dry polystyrene spheres were scattered over the grids and were shadowed with platinum at an angle of 8 to 1. The only segments of the DNA that were not covered with platinum were those that passed through the shadow of the large polystyrene spheres. These segments were located by tracing the path of the platinum-covered DNA strands until they entered the shadow area. High

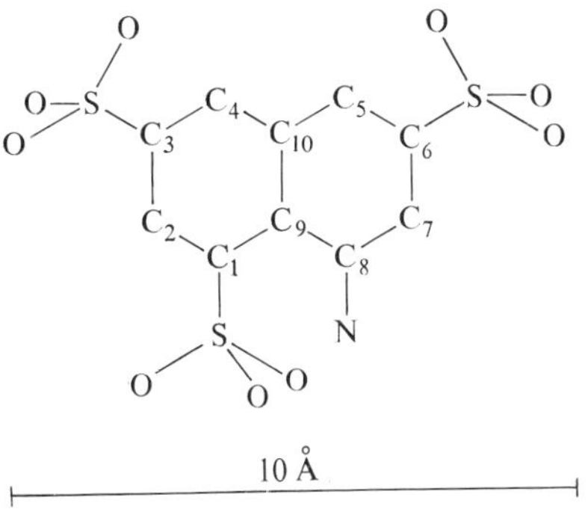

Fig. 9.2. Structure of 8-diazonium-1,3,6-naphthalenetrisulfonic acid. The maximum dimension spans the distance between two of the sulfonic acid groups.

resolution electron photomicrographs made of these areas showed that the DNA strand consisted of a row of dots in the shape of triangles 10 to 12 Å on a side, or strokes 10 to 12 Å long and 5 Å wide. The triangles were interpreted to be uranyl ions attached to each of the three sulfonic acid groups, while the strokes were seen as markers which lay in a plane perpendicular to the supporting film. It was estimated that three to six uranyl ions were attached to each marker molecule (one or two per sulfonate) and that this was sufficient to yield a detectable level of contrast over background.

One disadvantage of 8-diazonium-1,3,6-naphthalenetrisulfonic acid as a marker was its large size in relation to internucleotide distances. As shown in Fig. 9.2, the long diameter of this molecule is more than 10 Å as compared with a maximum separation of 7 to 8 Å between bases in a nucleic acid. Variations in the orientation of this marker in a nucleic acid, and problems of packing encountered in adjacent marked nucleotides would make this an unfavorable choice for precise distance measurements. Consequently, a more compact reagent, 2-diazonium-*p*-benzenedisulfonic acid was chosen for study as a marker (Moudrianakis and Beer, 1965a). A comparison was made of the rate of reaction of each of the four nucleotides with the diazonium reagent in order to assess its specificity. At pH 9 the reagent reacted with guanylic acid 60 times faster than with any other nucleotide or with the ribose portions

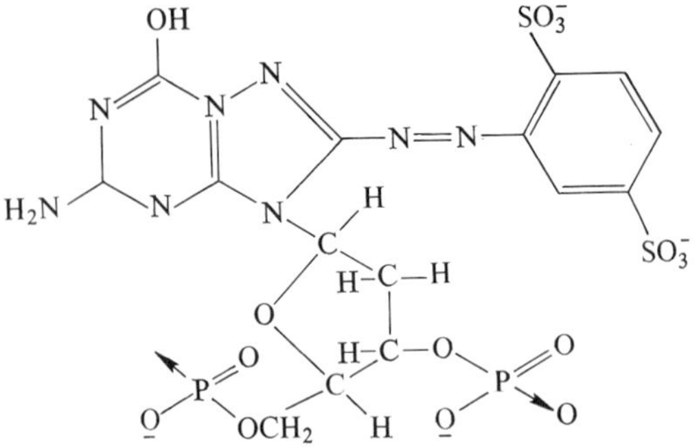

Fig. 9.3. Adduct of guanylic acid and diazotized 2-amino-*p*-benzenesulfonic acid.

of nucleic acid. The diazonium-guanylic acid adduct remained stable for several days. A proposed structure for this adduct is shown in Fig. 9.3.

Between two and four uranium atoms would presumably be bound by the two sulfonic acid groups, and it was assumed that this should still provide sufficient contrast to locate the marked bases. As shown in Fig. 9.3, the distance between the two sulfonic acid groups was estimated to be 10 to 12 Å (Moudrianakis and Beer, 1965b). This would also be too large unless the orientation was favorable. One such orientation is with the "stroke" connecting the marker groups perpendicular to the direction of the polynucleotide chain. Evidence that this was the case was provided by high resolution electron photomicrographs of marked and stained single-strand DNA (Moudrianakis and Beer, 1965b). When the DNA was followed into the shadow of a polystyrene sphere, the two dots constituting the marker were seen as generally perpendicular to the direction of the polynucleotide strand. The minimum distance between consecutive pairs of dots was found to be 6 to 7 Å. It was inferred that this marked the positions of consecutive guanines.

Under the conditions of this study, no excessive degradation of the polynucleotide chains was observed. However, no more than 80% of the guanine residues could be marked. If the unlabeled 20% represent the same sites in all the molecules, then a portion of the guanines will always remain undetected by this procedure. On the other hand, it is more likely that the unmarked guanines were distributed randomly, in which case the required data could be obtained from several identical molecules.

It would appear that all of the basic tools and techniques are available for the determination of sequences by electron microscopy. What are the prospects for carrying out such a study? One obvious approach would be to apply this procedure to a nucleic acid with a known sequence, preferably one that is single-stranded and devoid of any secondary structure. Suitable candidates may be found from among the various tRNAs and 5S RNAs. Under conditions of low ionic strength, the length of an extended tRNA chain may be estimated to be 500 Å and the length of 5S RNA as 800 Å. The location of a series of predictably spaced dots over a distance of 500 or 800 Å may not present insurmountable difficulties. For that matter, RNA segments of known sequence, replicated in the manner described in Chapter 8, may also yield suitable candidates for electron microscopic sequence determination. In addition to serving as an independent check of the other sequence methods, trials of this sort should provide the needed experience and build confidence in the reliability of electron microscopic sequence analysis.

Assuming that no further basic technical developments are needed, valuable information may also be obtained from the application of electron microscopic sequence analysis to viral nucleic acids such as ϕX-174 DNA and TMV RNA. In the latter case, the electron microscopic examination may be limited to small lengths through use of the stripping reaction described in Chapter 8. The partially stripped TMV would be marked with the diazonium benzene disulfonic acid, stained with uranyl acetate and examined with the electron microscope. The exposed nucleic acid could be located as a tail that protrudes from the end of a partially stripped virus. Thus the necessity for platinum shadowing would be avoided, and the sequence analysis could be performed on all of the nucleic acid that was visible. From the data reported previously (Mandeles, 1968; Lloyd and Mandeles, 1970), one could predict that two long segments of 180 and 490 Å containing no stained markers should be located near the right end of the TMV. These would represent the positions of the Ω-mer and Ψ_1-mer.

X-RAY CRYSTALLOGRAPHIC ANALYSIS

In recent times, X-ray crystallography has played a central role in the elucidation of protein and nucleic acid structure. With proteins, X-ray diffraction patterns have provided information mainly about conformation (see review by Streyer, 1968), but a few studies of high resolution patterns have furnished information about sequence also (Lipscomb *et al.*, 1968). With nucleic acids, the X-ray diffraction analyses of DNA by Wilkins *et al.* (1953) and Franklin and Gosling (1953) served as the basis for modern theory of DNA structure (Watson and Crick, 1953). Since then, the production of high quality diffraction patterns has resulted in refinements in our knowledge of DNA structure (Langridge *et al.*, 1957, 1960a, 1960b). A number of excellent reviews have been written on the application of X-ray crystallography to structural studies of biopolymers, including those of Holmes and Blow (1965) and Bullen (1962).

In a typical X-ray diffraction study, a crystal of the substance to be examined is placed in a narrow beam of X-rays and a photograph is made of the reflections as the crystal is moved in a regular way. Because the atoms of a crystal are arranged in a three-dimensional lattice, strong reflections in defined spots are produced as the crystal is moved. The spots that appear on the photographic plate correspond to specific directions and are related, reciprocally, to the spacing of the atoms in the crystal. Information from a series of X-ray diffraction patterns is used to reconstruct the positions of the reflecting atoms in maps or models. It is in the latter phase of the work that most of the effort is expended.

The various atoms (with the exception of hydrogens) begin to be distinguished from each other at a resolution of about 2 Å—well within the capability of available instrumentation. As a consequence, X-ray diffraction has often stimulated speculation about its eventual use in nucleic acid sequence analysis. These speculations have rarely been concerned with obtaining patterns from nucleic acid molecules larger than tRNA or 5S RNA because of the increase in pattern complexity associated with increased molecular size. An analysis of a typical tRNA alone would require information about the positions of approximately 1600 atoms, exclusive of the hydrogens. Nevertheless, a plan for the use of X-ray diffraction in the sequence analysis of tRNA might include the following stages:

1. Crystallization of a tRNA.
2. Determination of overall conformation by X-ray diffraction at a resolution level of about 10 Å.
3. Determination of the positions of purine and pyrimidine rings with respect to the ribose phosphate backbone, at about 4 Å resolution.
4. Discrimination of adenine from guanine at a resolution of about 2 Å. Cytosine and uracil will probably not be distinguished at this level because of the similarity in electron density of the amino nitrogen and the keto oxygen attached to the C-1 of cytosine and uracil, respectively.
5. Use of isomorphic "marked" tRNA to locate specific bases.
6. Correlation of the data into a sequence.

Although not necessarily directed to the ultimate goal of nucleic acid sequence analysis, progress has been made in the first two stages. Large hexagonal crystals of yeast formylemthionyl tRNA and *E. coli* phenylalanyl tRNA have been obtained by Hampel *et al.* (1968), and similar crystals have been obtained by Kim and Rich (1968). In the former study, clearly discernible diffraction maxima were seen at a 10 Å resolution. Likewise, Kim and Rich (1969) obtained full representation of diffraction data to a limit of 12 Å resolution and partially usable data down to a 7 Å level. The latter observations were made with orthorhombic crystals of *E. coli* formylmethionyl tRNA (Kim *et al.*, 1969). The data indicated that in the crystal the molecules are elongated and dimerized in an overlapping antiparallel fashion. The dimer length was estimated as 109 Å and the width 35 Å. The individual molecules were about 80 Å long, with an irregular cross section measuring 25 to 35 Å. By analogy with the experience in X-ray analysis of proteins, it is expected that the knowledge of the primary structure of nucleic acids will be helpful in establishing the three dimensional structure in crystals.

MEASUREMENT OF OPTICAL PROPERTIES IN SEQUENCE ANALYSIS

The ability to absorb UV light is probably the one physical property that is most widely used in the identification and determination of nucleic acids. This ability is (almost) wholly due to the interaction of the light with purine and pyrimidine chromophores. Because the absorption spectra of the individual bases are different, methods have been developed for the determination of nucleic acid base compositions in intact molecules which consider the relative contribution of each base (or nucleoside) to the total absorption spectrum (see Toal *et al.*, 1968; Richards, 1968). With respect to sequence analysis, the influence of permuted base sequences on conventional UV absorption spectra is too small to be measured in most cases. However, another property, optical activity, appears to be more sensitive to changes in sequence.

Ordinarily, the optical activity of a nucleic acid in the visible region is due to the structural asymmetry of the ribose moieties. With purines or pyrimidines attached to the C-1 of the ribose, strong optical activity is also observed in the UV region although the chromophores are themselves not optically active and the ribose does not absorb in this region. This phenomenon is ascribed to the perturbation of the symmetric base chromophores by the asymmetric field of the sugar. In addition, the bases are subject to (other) asymmetric conformational restrictions due to interaction with their nearest neighbor bases. As a result, nucleic acid solutions refract and absorb polarized UV light in an asymmetric fashion. The measurement of asymmetric refraction at a single wavelength is more familiarly known as optical rotation, and over a range of wavelengths it is known as optical rotatory dispersion (ORD). The measurement of asymmetric absorption over a range of wavelengths is known as circular dichroism (CD). Because ORD, a dispersion property, and CD, an absorption property, share a common origin, the optical activity, it is possible to calculate the dispersion properties of a system from its absorption properties and vice versa (Moffitt and Moscowitz, 1959; Moscowitz, 1960). Consequently, observations made with ORD are equally valid for CD. Thorough discussions of these optical properties as related to biopolymers may be found in reviews by Mommaerts (1968), Adler and Fasman (1968), Beychock (1968), Yang and Samejima (1969), and Tinoco and Cantor (1970).

The influence of base sequences on optical properties first becomes apparent in the ORD spectra of the 16 possible dinucleoside phosphates (Warshaw and Tinoco, 1965,1966). These are shown in Fig. 9.4 together with the ORD spectra of the four mononucleoside phosphates. It may be seen that the measured ORD spectrum of each dinucleoside phosphate differs from the composite spectrum of constituents. In addition, sequence isomers such as AU and UA, AC and CA, AG and GA, GC and CG, GU and UG, and CU and UC are easily distinguished. This is believed to be due to the more or less restricted orientation imposed by one base of the dimer upon the other. This tendency of nearest neighbor bases to assume a restricted orientation, called stacking, is apparently different for each of the four common found in nucleic acids. Stacking of nearest neighbor bases depends on a number of factors, including temperature, pH, and ionic strength. The effect of the order of the bases on the optical properties can be enhanced by judicious selection of these conditions.

While nearest neighbor base interaction serves as the basis for a sequence analysis tool, it also provides the fundamental limitation of the method since the strongest optical effects arise *only* from nearest neighbor interactions. The ORD and CD spectra of certain oligonucleotides may be difficult to distinguish for this reason. As an example, consider the sequence isomers CAUGA and GAUCA. The nearest neighbor interactions of the first oligonucleotide are: CA, AU, UG, and GA. For the second they are: GA, AU, UC, and CA. The only difference between the two is the UG in the former and UC in the latter. This difference may be too small to be discernible in an ORD or CD spectrum which includes the optical effects produced by the other base interactions.

Up to what chain length is ORD or CD responsive enough to permutations in base sequence to be used as an analytical tool? For trimers, the ORD spectra of certain oligonucleotides have been measured and compared with curves calculated from the ORD spectra of the constituent dimers and monomers (Cantor and Tinoco, 1967). The calculated curves agree remarkably well with the measured curves and provide support for the idea that calculated ORD spectra may be identical with the experimentally determined curves for all trimers. Though this is not strictly true (especially for trimers with U as the second base; Gray, 1969), the calculated curves do indicate that correct sequence assignments may still be made on the basis of ORD or CD spectral differences. This is especially the case when the base composition of the trimers is

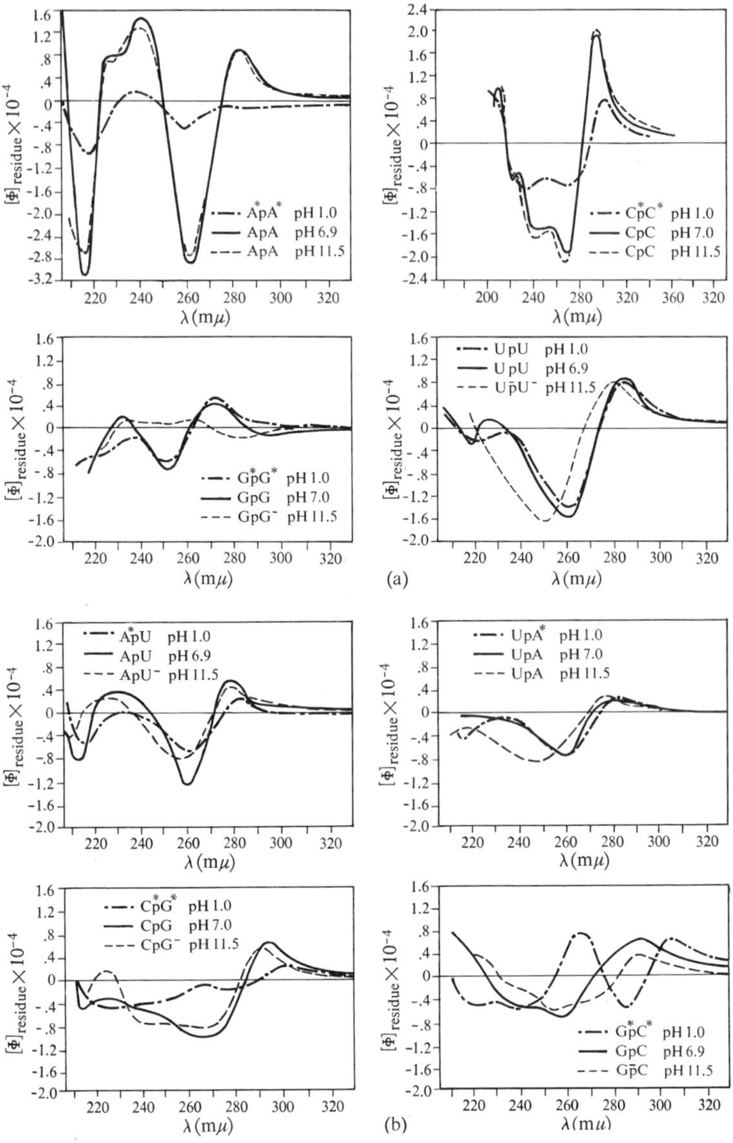

Fig. 9.4. Optical rotatory dispersion of 16 dinucleoside phosphates of RNA (from Sober, 1968).

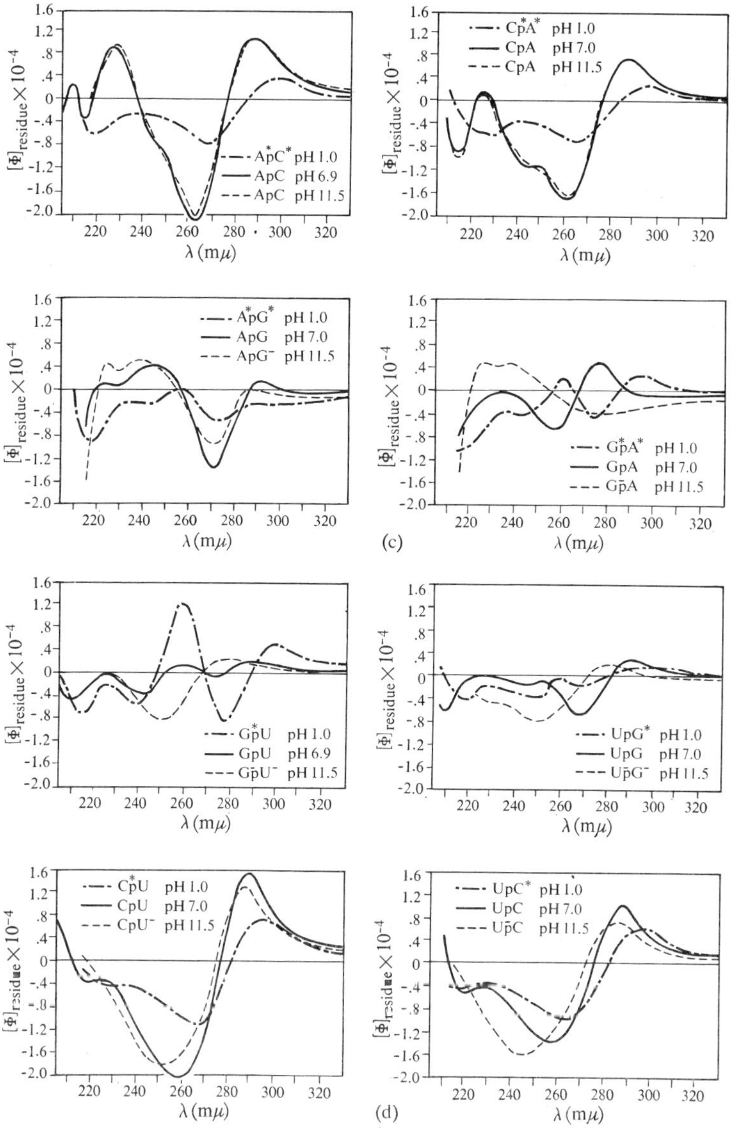

ApC* pH 1.0
ApC pH 6.9
ApC pH 11.5
CpA* pH 1.0
CpA pH 7.0
CpA pH 11.5
ApG* pH 1.0
ApG pH 7.0
ApG⁻ pH 11.5
GpA* pH 1.0
GpA pH 7.0
GpA pH 11.5
(c)
GpU pH 1.0
GpU pH 6.9
GpU⁻ pH 11.5
UpG* pH 1.0
UpG pH 7.0
UpG⁻ pH 11.5
CpU pH 1.0
CpU pH 7.0
CpU⁻ pH 11.5
UpC* pH 1.0
UpC pH 7.0
UpC pH 11.5
(d)
[Φ]residue × 10⁻⁴
λ (mμ)

known. As an example, the calculated ORD and CD spectra for the six possible sequence isomers of (AUG) are shown in Fig. 9.5. The curve for each oligonucleotide is sufficiently different from all of the rest so that even a mixture of the six could be quantitatively analyzed by computer methods (Tinoco and Cantor, 1970).

For tetramers and longer oligonucleotides, it is expected that differences in ORD spectra will not be so apparent as those in trimers or dimers for the reasons outlined above and because the spectra of the longer oligonucleotides will begin to approach those of nucleic acids with similar base compositions. However, the usefulness of ORD and CD may be extended to tetramers and possible pentamers by considering only those oligonucleotides that are produced by specific RNase action. As many as 27 different tetramers may be found in a T_1 RNase digest of RNA, and as many as 16 tetramers may be found in a pancreatic RNase digest. Given a tetramer fraction from a T_1 RNase digest with the base composition (ACGU), by analogy with the trimers, the task is reduced to distinguishing the sequence isomers AUCG, ACUG, CAUG, CUAG, UACG, and UCAG. The CD spectra calculated for these tetramers are shown in Fig. 9.6. As in the case of trimers, recognizable differences exist.

It would seem that all that is necessary for the sequence analysis of an unknown small oligonucleotide would be to match the ORD or CD spectrum of the unknown with the calculated curves of oligonucleotides of the same base composition. This has been accomplished in one instance of sequence analysis (Mandeles, 1967). In this study, TMV RNA was hydrolyzed with T_1 RNase and the right terminal oligonucleotide was isolated and analyzed for base composition and ORD. The oligonucleotide contained 3C and an A, and the ORD indicated that the most likely sequence was CpCpCpA. This was confirmed by SPDE hydrolysis of the terminal oligonucleotide which yielded 3Cp and an A as products. Apart from this one instance, the use of ORD or CD for sequence analysis of tetramers or smaller oligonucleotides has been largely overshadowed by the development of the fingerprint methods described in Chapter 4. It may be recalled that reliable sequence analyses may be made simply by partial digestion of an unknown oligonucleotide with SPDE, followed by ionophoresis on DEAE-cellulose. Although this is a destructive analysis, it requires only very small amounts of radioactive oligonucleotide.

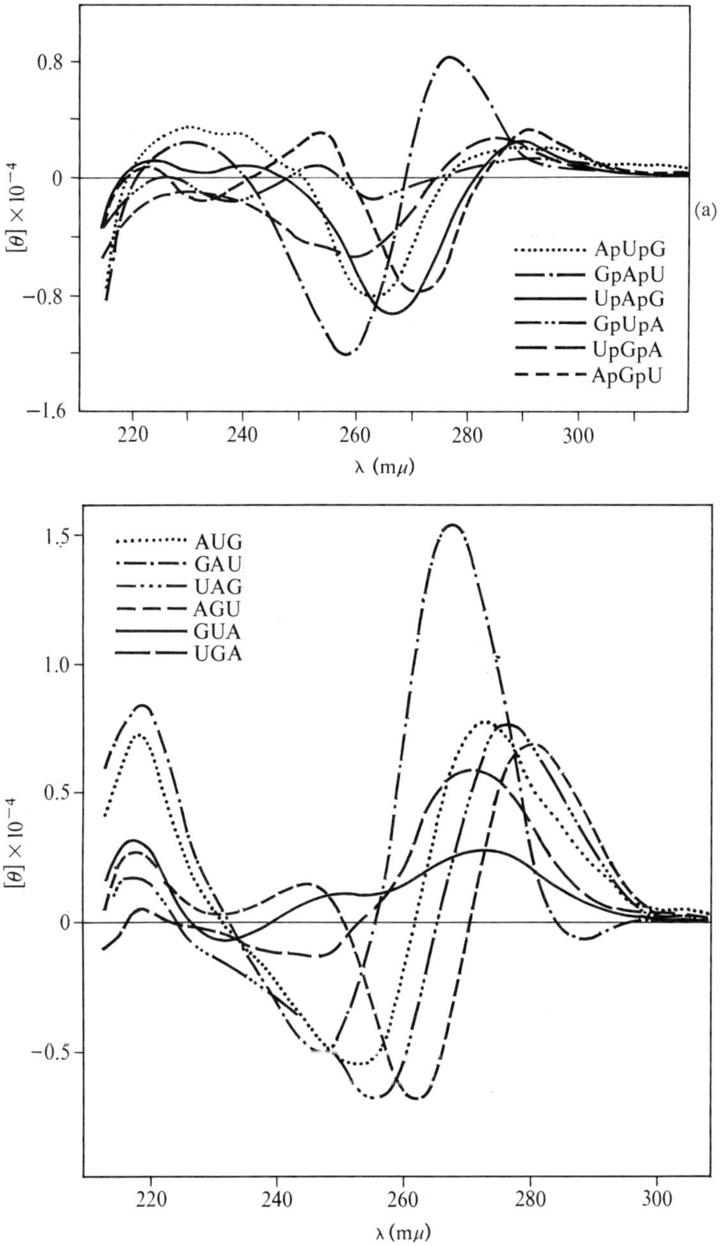

Fig. 9.5. Calculated optical properties of six sequence isomers with base composition AUG. (a) Optical rotatory dispersion (ORD) (from Cantor and Tinoco, 1967). (b) Circular dichroism (CD).

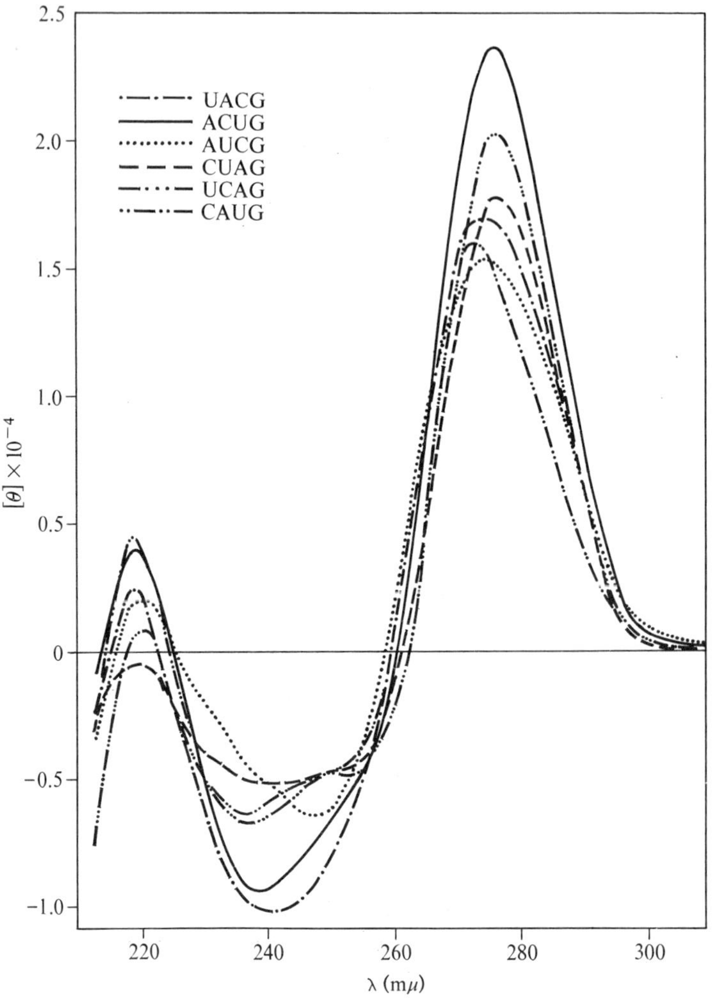

Fig. 9.6. Calculated CD spectra for tetramers. Calculations restricted to the base composition of (ACU)G in order to represent a product expected after T_1 RNase digestion and partial purification.

MASS SPECTROMETRY

The application of mass spectrometry in nucleic acid sequence analysis is probably restricted to the determination of dimers, trimers, and tetramers. This limitation on the basis of size stems mainly from the requirements for:

1. A volatile oligonucleotide or derivative.
2. Fragment ions that yield information on the order of the bases in the oligonucleotide.

For dimers, the first requirement is met by the use of trimethylsilylated (TMSO) derivatives (Hunt *et al.*, 1968). The second is met by the difference in susceptibility to fragmentation of the 3′-linked nucleotide from the 5′-linked nucleotide. As a result, the relative abundance of the nucleoside fragments serves as an accurate indicator of the sequence of the bases in the original dimer. Data concerning the fragment ions produced in the mass spectrometric analysis of silylated dimers have been collected for 13 of the 16 possible sequence isomers (Hunt *et al.*, 1968). The results show that base compositions and sequence may be determined on the basis of the mass and relative abundance of the ionized silylated nucleoside fragments.

The order of the bases may also be determined by the sequential use of two reagents to form derivatives of an oligonucleotide in an asymmetric fashion (Dolhun and Wiebers, 1969). For example, a dinucleoside monophosphate is first allowed to react with phenylboronic acid to form the diester of the free 2′- and 3′-hydroxyls of the nucleoside on the right end. Subsequently, TMSO groups are introduced into the remaining reactive positions to enhance the volatility of the dimer. In this manner, the nucleoside at the right end of an oligonucleotide is tagged with a boron and serves to distinguish it from the left terminal (or interior) nucleoside fragments. The use of this type of procedure for a trimer such as ApUpG is shown in a hypothetical simplified fragmentation diagram in Fig. 9.7. It may be seen that the right terminal nucleoside fragment contains the phenylboron ester, the left terminal fragment contains TMSO groups in the 2′- and 5′-hydroxyl positions, and the interior nucleoside contains only a single TMSO group in the 2′-hydroxyl position. Thus the identity and position of

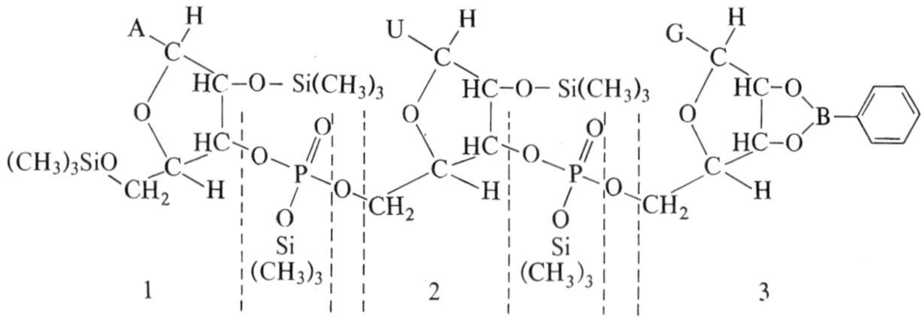

Fig. 9.7. Hypothetical fragmentation diagram for ApUpG. The dashed lines indicate the positions of the fragmentation points.

nucleoside fragments in a trimer may be determined. In the case of a tetramer such as CpApUpG, difficulty would be encountered in establishing the correct order of the internal adenosine and uridine derivatives unless:

1. The susceptibility to fragmentation was sequence dependent; or
2. if the mass spectrometry could be extended to the determination of the higher mass ionized dimer and trimer fragments from the fragmentation of the tetramer.

For small nucleic acids such as tRNA, low volatility together with the complexity of the fragments produced makes it unlikely that mass spectrometry will be used, even with automated acquisition of high resolution data and computer-assisted interpretation of spectra.

NUCLEAR MAGNETIC RESONANCE SPECTROSCOPY

In nuclear magnetic resonance (NMR) spectroscopy, the energy required to change the alignment of magnetic nuclei in a magnetic field is measured. Information of chemical interest arises from the fact that nuclei of atoms in different chemical environments are also generally in different magnetic environments, and come into resonance with a fixed-frequency oscillator at different values of an applied magnetic field. The principle of this method is illustrated in Fig. 9.8. The low-power radio transmitter operates at a fixed frequency such as

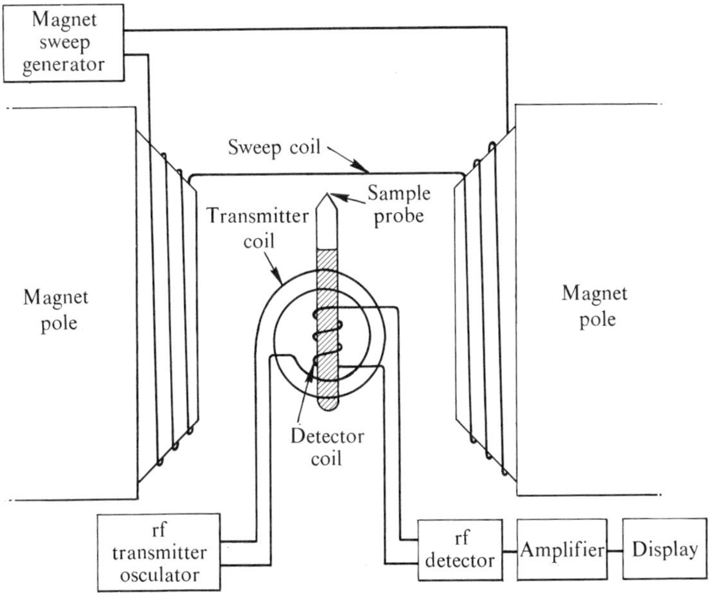

Fig. 9.8. Diagram of a conventional NMR apparatus.

40, 60, or 100 $\times 10^6$ hertz and causes a small oscillating electromagnetic field to be applied to the sample in the test tube. The magnetic field of the large magnet is held at a known value between 0 and 10,000 gauss, and the resonance condition is achieved by the superimposition of a small variable sweep field (38 milligauss, for example) on this larger fixed field (7000 gauss, for example). Voltage fluctuations in the receiving coil that occur as a result of the small sweep field may be amplified and recorded in an oscillograph tracing.

Magnetic properties are found with nuclei of odd-numbered masses, such as ^1H, ^{13}C, ^{15}N, ^{17}O, ^{19}F, and ^{31}P, and nuclei of even mass but odd atomic number, such as ^2H, ^{10}B, ^{14}N. Nuclei such as ^{12}C, ^{16}O, ^{32}S with both even mass and atomic number, have no magnetic properties and thus do not yield magnetic resonance signals. Detailed discussions of the principles and techniques of NMR spectroscopy may be found in numerous textbooks. The applications of NMR to biology have been described in an extensive review by Jardetzky and Jardetzky (1962).

With respect to nucleic acids, interest has centered mainly on the magnetic properties of the hydrogens—proton magnetic resonance (PMR). In two elegant studies by Ts'o and his collaborators (Ts'o *et al.*, 1969, 1970), oligonucleotides such as dimers and trimers were examined by PMR spectroscopy. The spectra revealed that there was a very strong interaction between the magnetic fields of nearest neighbor bases, and from this, information could be deduced not only about sequence but also about conformation. As with ORD and CD, the usefulness of PMR is probably limited to the identification of oligonucleotides of the size of trimers or tetramers.

REFERENCES

Adler, A. J., and Fasman, G. D. (1968). In *Methods in Enzymology*, Vol. 12B, L. Grossman and K. Moldave (eds.). Academic Press, New York.

Beer, M., and Moudrianakis, E. N. (1962). *Proc. Natl. Acad. Sci.*, **48**, 409.

Beer, M., and Zobel, C. R. (1961). *J. Mol. Biol.*, **3**, 717.

Beychock, S. (1968). *Ann. Rev. Biochem.*, **37**, 437.

Bullen, G. J. (1962). In *Comprehensive Biochemistry*. M. Florkin and E. H. Stotz (eds.). Elsevier Publishing Co., Amsterdam.

Cantor, C. R., and Tinoco, I., Jr. (1967). *Biopolymers*, **5**, 821.

Cantor, C. R., Tinoco, I., Jr., and Peller, L. (1964). *Biopolymers*, **2**, 51.

Dolhun, J. J., and Weibers, J. L. (1969). *J. Am. Chem. Soc.*, **91**, 7755.

Franklin, R. E., and Gosling, R. G. (1953). *Nature*, **171**, 785.

Gray, D. (1969). Personal communication.

Griffith, J. (1969). Ph.D. Thesis, California Inst. Tech., Pasadena, California.

Hampel, A., Labanauskas, M., Connors, P. G., Kirkegard, L., RajBhandary, U. L., Sigler, P. B., and Bock, R. M. (1968). *Science*, **162**, 1384.

Holmes, K. C., and Blow, D. M. (1965). *Methods Biochem. Anal.*, **13**, 113.

Hunt, D. F., Hignite, C. E., and Biemann, K. (1968). *Biochem. Biophys. Res. Commun.*, **33**, 378.

Jardetzky, O., and Jardetzky, C. D. (1962). *Methods Biochem. Anal.*, **9**, 235.

Kim, S.-H., and Rich, A. (1968). *Science*, **162**, 1381.

Kim, S.-H., and Rich, A. (1969). *Science*, **166**, 1621.

Kim. S.-H., Schofield, P., and Rich, A. (1969). *Cold Spring Harbor Symp. Quant. Biol.*, **34**, 153.

Langridge, R., Seeds, W. E., Wilson, H. R., Hooper, C. W., Wilkins, M. H. F., and Hamilton, L. D. (1957). *J. Biophys. Biochem. Cytol.*, **3**, 767.

Langridge, R., Wilson, H. R., Hooper, C. W., Wilkins, M. H. F., and Hamilton, L. D. (1960a). *J. Mol. Biol.*, **2**, 19.

Langridge, R., Wilson, H. R., Hooper, C. W., Wilkins, M. H. F., and Hamilton, L. D. (1960b). *J. Mol. Biol.*, **2**, 38.

Lipscomb, W. N., Hartsuck, J. A., Reeke, G. N., Jr., Quiocho, F. A., Bethge, P. H., Ludwig, M. L., Steitz, T. A., Muirhead, H., and Coppola, J. C. (1968). *Brookhaven Symp. Biol.*, **21**, 24.

Lloyd, D. A., and Mandeles, S. (1970). *Biochemistry*, **9**, 932.
Mandeles, S. (1967). *J. Biol. Chem.*, **242**, 3103.
Mandeles, S. (1968). *J. Biol. Chem.*, **243**, 3671.
Moffitt, W., and Moscowitz, A. (1959). *J. Chem. Phys.*, **30**, 3.
Mommaerts, W. H. F. M. (1968). In *Methods in Enzymology*, Vol. 12B. L. Grossman and K. Moldave (eds.). Academic Press, New York.
Moscowitz, A. (1960). In *Optical Rotatory Dispersion*. C. Djerrassi (ed.). McGraw-Hill, New York.
Moudrianakis, E. N., and Beer, M. (1965a). *Biochim. Biophys. Acta*, **95**, 23.
Moudrianakis, E. N., and Beer, M. (1965b). *Proc. Natl. Acad. Sci.*, **53**, 564.
Richards, E. G. (1968). *Eur. J. Biochem.*, **4**, 256.
Sober, H. A. (ed.) (1968). *Handbook of Biochemsitry*. The Chemical Rubber Co., Cleveland, Ohio.
Streyer, L. (1968). *Ann. Rev. Biochem.*, **37**, 25.
Tinoco, I., Jr., and Cantor, C. R. (1970). *Methods Biochem. Anal.*, **18**, 81.
Toal, J. N., Rushizky, G. W., Pratt, A. W., and Sober, H. A. (1968). *Anal. Biochem.* **23**, 60.
Ts'o, P. O. P., Kondo, N., Schweizer, M. P., and Hollis, D. P. (1969). *Biochemistry*, **8**, 997.
Ts'o, P. O. P., Schweizer, M. P., and Hollis, D. P. (1969). *Ann. N.Y. Acad. Sci.*, **158**, 256.
Warshaw, M. M., and Tinoco, I., Jr. (1965). *J. Mol. Biol.*, **13**, 54.
Warshaw, M. W., and Tinoco, I., Jr. (1966). *J. Mol. Biol.*, **20**, 29.
Watson, J. D., and Crick, F. H. C. (1953). *Nature*, **171**, 737.
Wilkins, M. H. F., Stokes, A. R., and Wilson, H. R. (1953). *Nature*, **171**, 780.
Yang, J. T., and Samejima, T. (1969). *Prog. Nucleic Acid Res.*, **9**, 223.
Zobel, C. R., and Beer, M. (1961). *J. Biophys. Biochem. Cytol.*, **10**, 333.

10.

ISOLATION OF PURIFIED NUCLEIC ACID

Up to now we have been concerned with the different strategies used in nucleic acid sequence analysis. The wide variety of experimental techniques that have been used to implement these strategies attest to the intense interest in this area of research. Fortunately, a number of reviews and experimental exegeses have been published that embody some of the best features of the various techniques and procedures (Chargaff and Davidson (1955, 1960); Colowick and Kaplan, 1960; Grossman and Moldave, 1967, 1968; Cantoni and Davies, 1966). In this discussion, liberal use will be made of these earlier experimental descriptions as indeed they are liberally used in many laboratories.

The procedures used in nucleic acid sequence analysis may be divided into two groups. The first is concerned with the isolation of molecularly intact, homogeneous nucleic acid; the second comprises the experimental determination of primary structure. The first group of techniques is the subject of this chapter, and the second group is treated in Chapter 11.

Because the trend in nucleic acid sequence analysis is moving toward the use of radioactive nucleic acids (owing in large part to the development of the fingerprint analytical method of Sanger *et al.*, 1965), the following procedures are illustrated with ^{32}P. This isotope may be obtained with a high specific activity and, because phosphorus is an integral part of the repeating unit in the nucleic acid backbone, it helps distinguish nucleic acids from any other biopolymer.

GENERAL CONSIDERATIONS

There are three main stages in the isolation of purified radioactive nucleic acids: the growth of an organism in the presence of radioactive label, the separation of a crude nucleic acid fraction from the other cellular components, and the isolation of a specific nucleic acid suitable for sequence analysis. The first stage is usually accomplished under nearly the same conditions that are used for normal cell growth. The only variations occur when an attempt is made to restrict the supply of an element that is to be administered as the radioactive isotope. For example, ^{32}P-labeled cells are often grown in a medium containing a minimum of nonisotopic phosphate.

The procedures used in the third stage are essentially the same for any one type of nucleic acid: the isolation of a specific tRNA is the same whether or not the tRNA mixture originated in yeast, bacteria, plant, or mammalian cells. Similarly, the isolation of rRNA from ribosomes is essentially the same regardless of the source of the ribosomes.

In terms of experimental procedures, the major differences occur in the second stage, where the nature of the other cellular constituents governs the details of the procedures to be used. In mammalian tissues, for instance, the presence of a mixture of cell types, connective tissue, fat depots, etc., tends to increase the difficulties in the isolation of specific nucleic acids. More steps are required to isolate specific nucleic acids from histologically complex mammalian tissues than from a culture of bacteria. Because of this, the source of mammalian cells is usually cell cultures. The details of cell and tissue culture techniques are beyond the scope of this volume. Valuable information may be obtained from a number of recent experimental descriptions (Porterfield, 1967; Bachrach and Breese, 1968; Holper, 1969; Penman *et al.*, 1969).

The following directions are for the isolation of ^{32}P-labeled tRNA, 5S RNA and rRNA from *E. coli*, and rabbit reticulocytes, mRNA from rabbit reticulocytes, bacteriophage R17 RNA from infected *E. coli*, and TMV RNA from infected tobacco. These particular procedures were chosen because they illustrate experimental manipulations that are common to the preparation of nucleic acids from other sources such as yeast and plants (see Chapters 4 and 6 in Grossman and Moldave, 1967, 1968).

GROWTH OF ^{32}P-LABELED E. coli

Conditions are given for the growth of low nuclease strains of *E. coli.* For other organisms, the main requirement is for a medium that contains the minimum amount of phosphate and that this amount be composed of ^{32}P. The medium used for the *E. coli* strains A19, Q13, and MRE600 is essentially that of Garen and Levinthal (1960) except for a decrease in the amount of sodium glycerphosphate.

GROWTH MEDIUM

NaCl	8×10^{-2} *M*
$FeCl_3$	2×10^{-6} *M*
KCl	2×10^{-2} *M*
Tris buffer, pH 7.5	1.2×10^{-1} *M*
Sodium glycerophosphate	10^{-4} *M*
$CaCl_2$	2×10^{-4} *M*
NH_4Cl	2×10^{-2} *M*
$MgCl_2$	10^{-3} *M*
Na_2SO_4	5×10^{-4} *M*
Glucose	1.2×10^{-2} *M*
Bactopeptone (Difco)	0.04%
^{32}P-phosphoric acid	1–20 mC

Enough medium should be made to provide for 1000 ml of culture and 10 ml of subculture. The subculture is inoculated with one of the strains of *E. coli* and incubated at 37° overnight (12 to 18 hours). The total volume of subculture is then used to inoculate the main culture medium in a 2000-ml flask. A parallel culture may be made without ^{32}P and the optical density of this read as a measure of the extent of growth in the flask containing ^{32}P. The culture is aerated vigorously at 37°, and the incubation is continued until the optical density of samples from the parallel culture reaches 0.5 at 550 mμ. This indicates that four to five generations of cells have been produced and that the cells are still in the exponential growth phase. (This should take no more than

5 hours of incubation under these conditions). The culture is chilled and centrifuged at 17,000 × *g* for 30 minutes. The supernatant is discarded and the cells are suspended in 30 ml of 0.01 *M* phosphate buffer, pH 7.5, and centrifuged again at 17,000 × *g* for 30 minutes. The cell pellet is retained for subsequent isolation of RNA.

EXTRACTION OF RNA FRACTIONS FROM ^{32}P.-LABELED E. coli (HOSOKOWA, 1970)

Summary of the Procedure. The radioactive *E. coli* is ground with alumina and the total RNA is freed from the cellular debris by centrifugation. The RNA extract is then centrifuged at high speed to separate ribosomes from tRNA. The ribosomes are separated into the 50S and 30S subunits. Extraction of the 50S subunits with phenol yields 23S RNA and 5S RNA, extraction of the 30S subunit yields 16S RNA.

Materials. Alumina for grinding is reagent grade obtained from Norton. The enzyme DNase (RNase-free) is obtained from Worthington Biochemicals, Inc. Buffers are prepared as follows:

TMA(−2) = 0.01 *M* Tris, 0.03 *M*
NH_4Cl, 0.01 *M* Mg acetate, 0.006 *M* β-mercaptoethanol, ph 7.8
TMA(−3) = 0.01 *M* Tris, 0.03 *M* NH_4Cl, 0.001 *M* $MgCl_2$, pH 7.5
TMA(−4) = same composition as TMA(−2) except that the Mg acetate concentration is 0.0003 *M*.

The ^{32}P-labeled *E. coli* cells harvested from 1 liter of culture (approximately 1 g wet weight) are placed in a chilled mortar. To this are added 2.5 g of alumina and 0.1 ml of solution containing 100 μg of DNase. The mixture is ground for about 10 to 15 minutes until the thick powdery mixture becomes a liquid paste. The paste is then diluted with 5 ml of TMA(−2) buffer and the grinding is continued for another 5 minutes. The heavy suspension is centrifuged at 17,000 × *g* for 15 minutes and the residue is discarded. The supernatant is then centrifuged in a No. 40 rotor for 4 hours at 37,000 rpm. The ribosomal pellet is reserved for the isolation of 23S, 16S, and 5S RNAs. The supernatant is used for the isolation of tRNA.

Isolation of Crude tRNA. The tRNA fraction from the previous step is mixed with 0.1 volume of 1 *M* sodium acetate buffer, pH 5.0, and 0.1 ml of bentonite suspension (containing 200 μg) (Fraenkel-Conrat *et al.*, 1961). This is extracted with an equal volume of phenol saturated with the same buffer. The extraction is performed with rapid stirring or shaking at room temperature for 15 minutes. Separation of the aqueous phase from the phenol phase may be hastened by centrifugation of the mixture at 2000 rpm for 10 minutes. The phenol phase is discarded and the aqueous layer is extracted again with fresh phenol saturated with acetate buffer. After separation of the phases, the aqueous layer is recovered and extracted three times with ether. The excess ether is removed under reduced pressure with the aid of a water aspirator. The aqueous solution is then mixed with 2.5 volumes of 95% ethanol chilled to $-20°$ and the mixture is allowed to stand at $-20°$ for at least an hour. The precipitate is collected by centrifugation at $17{,}000 \times g$ for 30 minutes and washed once with cold 95% ethanol.

The crude ^{32}P-tRNA may be further purified by either of the two following procedures:

(a) Polyacrylamide gel electrophoresis.
(b) Sucrose density gradient centrifugation.

POLYACRYLAMIDE GEL ELECTROPHORESIS OF tRNA EXTRACT

Special Equipment. Commercial disc electrophoresis, or flat slab electrophoresis equipment or an equivalent that is constructed of clear plastic, and power supply. The following directions are given for disc electrophoresis. A generalized illustration of an apparatus of this kind is shown in Fig. 10.1.

Separation of polynucleotides in gels is accomplished on the basis of differences in molecular size and shape as well as in mobility. The isolation of tRNA from a crude RNA extract depends on the use of a gel that impedes the movement of the higher molecular weight RNAs.

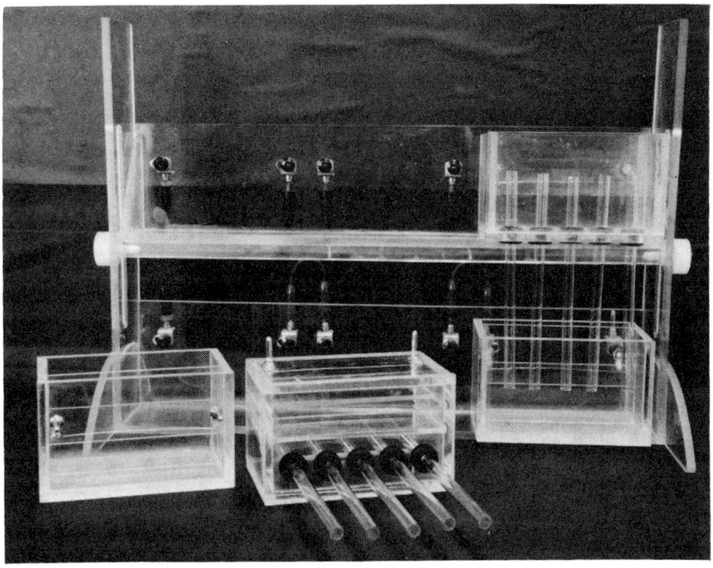

Fig. 10.1. Disc electrophoresis apparatus. Shown are the upper and lower reservoirs before assembly (on the left) and in place after formation of the gels in the gel tubes (on the right).

For this reason, the standard system of Richards *et al.* (1965) is used: a discontinuous buffer system with a 5% spacer gel and a 10% running gel. Modifications have been made in the buffer system to provide a pH of 8.2 rather than 8.9.

The buffers are prepared as follows:

Buffer A	48 ml 2 *M* HCl 20 g Tris base Dilute to 200 ml
Buffer B	48 ml 2 *M* HCl 26.3 g BisTris base (General Biochemicals) Water to 200 ml
Buffer C	10.75 g tricine (Calbiochem) 6.2 g Tris base Water to make 1 liter

Buffer A, after dilution, is used for the running gel and the lower electrode chamber (anode). Buffer B, after dilution, is used for the spacer gel and sample. Buffer C is used undiluted for the upper reservoir and electrode chamber (cathode).

The gel solutions are prepared as follows for the 10% running gel:

Solution 1 40 g acrylamide (Bio·Rad Laboratories)
0.6 g bisacrylamide
15 mg $K_3Fe(CN)_6$
Water to make 100 ml

Solution 2 50 ml buffer A
0.23 ml TEMED (NNN′N′-tetramethylethylenediamine)
50 ml water

Solution 3 0.14 g ammonium persulfate
Water to make 100 ml

This solution is degassed with the aid of a water aspirator pump.

Solution 4 Mix two parts of solution 1 with one part of solution 2 and degas = solution 4).

Mix equal parts of solution 3 and solution 4 and use to load polyacrylic tubes prepared as follows:

Polyacrylic plastic tubing, ⅜ inch OD and ¼ inch ID, is cut into 15-cm lengths. A 1-inch square of dialysis membrane is prepared by soaking it in water for 1 minute. The dialysis square is then placed over one end of the plastic tube and a ¼-inch length of ¼-inch ID amber rubber tubing is forced over that end to hold the membrane in place. The surface of the membrane should be flat and taut against the rim of the plastic tube. The running gel solution is poured into the tube to a height of 11 cm. A 1-cm layer of water is placed carefully on top of the gel solution and the tubes are held in a vertical position.

The polymerization should be essentially complete after a few minutes at room temperature. When the gels are set, the water over the gel surface is carefully removed and discarded.

The 5% spacer gel solutions are prepared as follows:

Solution 1 20.0 g acrylamide
0.6 g bisacrylamide
15 mg $K_3Fe(CN)_6$
Water to make 100 ml

Solution 2 50 ml Buffer B
0.23 ml TEMED
50 ml water

Solution 3 0.14 g ammonium persulfate
Water to make 100 ml
This solution is degassed with the aid of a water aspirator pump.

Solution 4 Mix two parts of solution 1 with one part of solution 2 and degas.

Mix equal parts of solutions 3 and 4 and use as spacer gel solution. Spacer gel solution is added to the surface of the running gel with a transfer pipet. Enough of the gel solution is added to reach a height of 0.5 to 1 cm above the surface of the running gel. Above this is placed some degassed water to a height of an additional centimeter, and the gel is allowed to polymerize without disturbance in a vertical position. Polymerization should be complete within 20 minutes. After the polymerization, the water layer is removed and discarded.

The tube containing the gel is inserted from the underside, through the o-ring seals, into the upper reservoir so that about 3 cm of the tube protrudes into the chamber. The lower reservoir, containing buffer A diluted 1:8, is placed directly below and the upper reservoir is lowered until the plastic tube is submerged to a depth of about 2 cm below the surface of the buffer. Any air bubbles adhering to the underside of the tube are removed with a pipet or with a syringe fitted with a length of small-bore teflon or polyethylene tubing.

The tRNA sample is mixed with an equal volume of 10% sucrose: this sucrose solution is made up in buffer B that has been diluted 1:4 and contains a small amount (0.2%) of brom phenol blue. The combined tRNA and sucrose solution is layered onto the surface of the spacer gel. The remainder of the tube is filled carefully with buffer C and then the upper reservoir is filled with buffer C. Care is exercised so that the layer of buffer above the sample is not disturbed during the filling of the upper reservoir. The negative electrode (cathode) is attached to the upper reservoir, and the positive electrode is attached to the bottom reservoir.

The electrophoresis may be carried out in a cold room with currents up to 5 mA per gel and voltages up to 100 volts. The electrophoresis is allowed to continue until the brom phenol blue band reaches a point above the bottom of the tube. After the run is over, the upper reservoir is emptied, the electrophoresis tubes are removed, and the membranes are removed from the bottoms. For visualization by staining, the gels are extruded into a test tube with the aid of a syringe filled with water. The test tube is filled with 10% acetic acid and the gel is allowed to stand at room temperature for 30 minutes. The acetic acid is replaced by a solution of 0.2% azure A dye in 0.4 *M* Na acetate buffer, pH 4.63, and the gel is allowed to stand for an hour at room temperature. At the end of this period, the excess dye is removed electrophoretically in an apparatus that is available commercially or by continuous washing with water (an overnight wash is usually sufficient).

The tRNA may be visualized as a comparatively wide band that migrates well away from the origin. It is distinguishable from the ribosomal RNA at or near the origin and other RNA fragments present in smaller bands.

LOCATION AND EXTRACTION OF MIXED ^{32}P-tRNA FRACTION

For the location and extraction of ^{32}P-tRNA, the gel is extruded onto a 10-inch length of ½ × ½ inch angle aluminum. This is placed in a tray containing powdered Dry Ice, and more powdered Dry Ice is used to cover the polyacrylamide gel. When the gel is completely frozen (this takes about 5 minutes), it is sliced into 1-mm fractions. Each disc is placed in the center of a stainless steel planchet and counted on a low background gas flow counter. As with the staining procedure, the tRNA is recognized as a comparatively wide band of radioactivity that migrates well away from the origin.

Using a glass homogenizer, the discs comprising the tRNA fraction are combined and homogenized in 1 ml of 0.05 *M* sodium acetate buffer, pH 5, containing 100 μg of bentonite. An equal volume of phenol saturated with 0.05 *M* sodium acetate buffer, pH 5, is added to the polyacrylamide homogenate and the mixture is shaken or stirred vigorously for 15 minutes at room temperature. The mixture is centrifuged at 2000 rpm to separate the two layers, and the RNA is precipitated from the aqueous phase by the addition of 2.5 volumes of cold 95% ethanol. This mixture is allowed to stand for several hours, and then the tRNA is recovered by centrifugation at 17,000 × *g* for 30 minutes. The precipitate is washed once with cold 95% ethanol and may be stored at −20°.

SUCROSE GRADIENT CENTRIFUGATION OF tRNA EXTRACT

Special Equipment. The special requirements include a Spinco Model L centrifuge or equivalent, SW 39 rotor or equivalent, Lusteroid tubes for rotor, 5-ml capacity gradient maker, gradient fractionator. The gradient maker may be a commercial model or may be made simply from a block of clear plastic as shown in Fig. 10.2. The collection

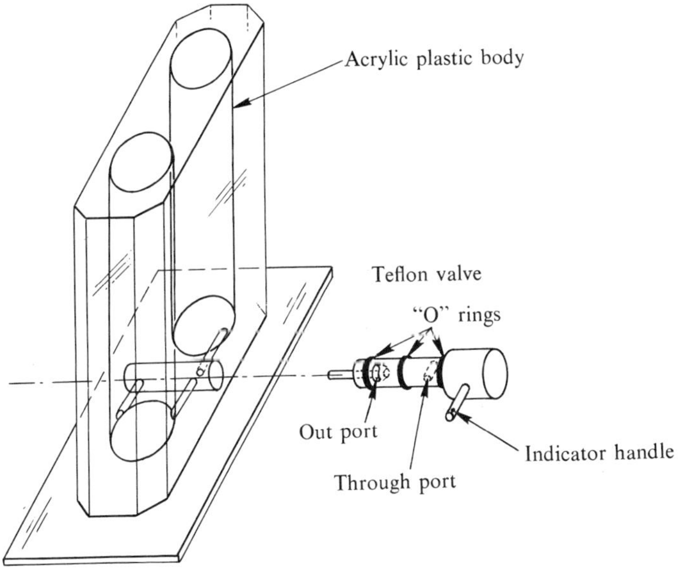

Fig. 10.2. Gradient maker.

of gradient fractions may be made by hand and gravity flow, or with the aid of a pump, UV monitor, and automatic fraction collector fitted with a drop counter.

Reagent grade of sucrose is used. If the presence of nuclease is suspected, treatment with bentonite will be necessary. This is accomplished by sedimenting 10 mg of bentonite or Macaloid through 25 ml of 5% or 20% sucrose at 28,000 rpm in a No. 30 Spinco rotor for 3 hours. Two sucrose solutions are made containing, respectively, 5 and 20% sucrose in 0.1 *M* sodium acetate buffer, pH 5. The sucrose solutions are loaded into a gradient maker as follows:

1. The valve between the chambers is shut.
2. The outlet valve is shut.
3. Twenty-four milliliters of the lighter sucrose solution (the 5% sucrose) are added to the reservoir chamber.
4. The valve connecting the two chambers is opened slightly to permit the 5% sucrose solution to fill the interconnecting channel. Any sucrose solution spilling into the mixing chamber may be returned to the reservoir with a transfer pipet.
5. Two and two-tenths milliliters of the denser solution (20% sucrose) are added to the mixing chamber (which is connected directly with the outlet tube).
6. The outlet tube is placed with its open end touching the inside surface of the Lusteroid tube at a point that is just a few millimeters higher than the level of the completed gradient.
7. The valve connecting the two chambers is opened cautiously, the stirrer in the mixing chamber is started, and the outlet valve is opened. If the sucrose solution fails to flow immediately, the valve between the two chambers is shut and slight pressure is applied to the mixing chamber with the heel of the hand. As soon as flow is resumed, the interconnecting valve between the reservoir and mixing chambers is opened.
8. The flow rate is adjusted so that it takes 15 minutes to complete the gradient in the centrifuge tube.

The purpose of the unequal volumes is to compensate for the differences in hydrostatic pressure of the two solutions.

9. The completed gradients are placed carefully into the buckets of the rotor and allowed to stand in the cold for 1 or 2 hours.

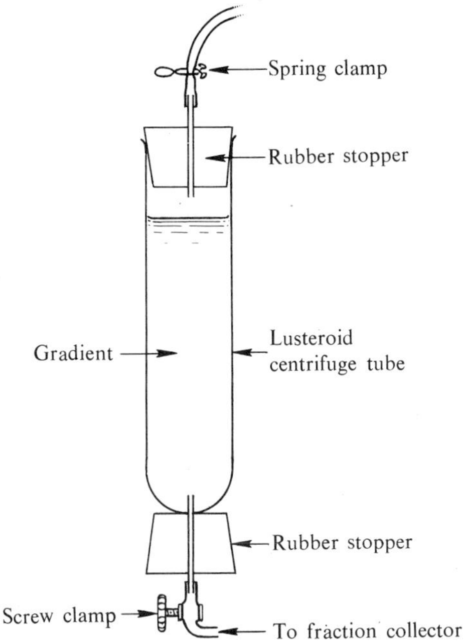

Fig. 10.3. Simple gradient collection device.

The ^{32}P-tRNA in 0.1 ml of 0.1 *M* sodium acetate buffer, pH 5, is loaded onto the gradient. A long transfer pipet is used to bring the sample in contact with the gradient surface and not the side of the tube. Use of any nearby solid support for the hand during this transfer is recommended. The caps of the buckets are coated with a thin film of high-vacuum grease and the caps are screwed into place. The buckets are secured and centrifugation is carried out at 39,000 rpm for 6 hours at 4°.

At the conclusion of the centrifugation, the sucrose gradient may be fractionated with the aid of a commercial device or samples may be collected manually (and inexpensively) as illustrated in Fig. 10.3. A one-hole rubber stopper that just fits into the mouth of the centrifuge tube is fitted with a short length of glass tubing to which a length of rubber tubing is attached. This provides a means of applying pressure to the gradient by blowing air into the tube, or stopping the flow out of the centrifuge tube by clamping the rubber tube shut. A second rubber stopper, a No. 6, is pierced through the center (axially) with a 2-inch needle of 22 gauge so that the point protrudes about 2 to 3 mm above the surface. The syringe connection is cut off and a length of Tygon tubing is attached to the end. A spring clamp is mounted on the rubber tubing attached to the stopper that fits the mouth of the centrifuge tube, and a small screw clamp is mounted on the Tygon tube attached to the needle in the small stopper. As shown in Fig. 10.3, the lower stopper is clamped firmly and supported by a ring stand, with the point of the needle projecting upward. The procedure for fractionation is as follows.

1. The rubber stopper is pressed firmly into the centrifuge tube with the spring clamp open, and the tube is mounted in a clamp attached to the ring stand. This clamp is loosened on the ring stand slightly and moved downward smoothly to force the bottom of the centrifuge tube onto the needle protruding from the lower rubber stopper. The centrifuge tube is pressed firmly onto the lower rubber stopper and secured in place. This should prevent leakage from the juncture of the centrifuge tube and the rubber stopper. If any leakage occurs, fractionation may be halted by shutting the clamp on the upper tube. The leak may then be repaired, usually by application of a small amount of vacuum grease.

2. With the upper clamp open, samples are collected by opening the clamp on the Tygon tube. The flow rate should be adjusted so that the entire gradient may be collected in 15 minutes.

The number of drops collected per fraction depends on the total gradient volume as well as the volume of each drop. It is desirable to collect between 30 and 50 fractions for optimal resolution of sedimented nucleic acids. A sedimentation pattern may be drawn from assays of the radioactivity of 1-μl portions of each fraction. The tRNA is recognized as the third or slowest sedimenting peak. Its position may be checked against that of an authentic sample of nonisotopic tRNA. The fractions corresponding to this peak are combined and the sucrose removed by dialysis against water or buffer.

PURIFICATION OF INDIVIDUAL ^{32}P-tRNAs

Much of the recent work on the isolation of specific tRNAs has relied on the use of at least two column chromatographic steps. The columns that appear to provide the widest range of application are: DEAE-Sephadex (Kawade *et al.*, 1963; Cherayil and Bock, 1965; Miyazaki and Takemura, 1966; Seno *et al.*, 1968), benzoylated DEAE-cellulose (BD-cellulose) Gillam *et al.*, 1967; Seno *et al.*, 1968), and reverse-phase partition chromatography with Chromosorb-W (Tanaka *et al.*, 1962; Muench and Berg, 1966; Kelmers *et al.*, 1965; Kelmers, 1966; Weiss *et al.*, 1968; Nishimura *et al.*, 1967). A typical procedure calls for separation of the crude tRNA mixture on DEAE-Sephadex and isolation of a fraction containing a specific amino acid-accepting activity. The fraction is then further purified by reverse-phase partition chromatography (Nishimura *et al.*, 1967) or by BD-cellulose chromatography (Seno *et al.*, 1968) and the purified tRNA again located by its amino acid-accepting ability. These methods have been developed for the isolation of comparatively large quantities of tRNA, while the trend in recent nucleic acid sequence analysis work is toward the use of small quantities of very radioactive tRNA. The assay using labeled amino acids for amino acid-accepting ability of a small quantity of ^{32}P, ^{14}C-, or ^{3}H-labeled tRNA is often difficult to carry out, especially in the later stages of purification. The success of the column methods may therefore depend on establishing the elution position of specific tRNAs from the chromatography and assay of larger quantities of nonradioactive material.

A more direct approach to the isolation of specific tRNAs has been devised by Gillam *et al.* (1968). Although the procedure as reported is suitable for the purification of large quantities of specific tRNAs, it has been scaled down for the isolation of 95% pure ^{32}P-$tRNA_1^{val}$ from *E. coli* (Yaniv and Barrell, 1969). In this procedure, the affinity of BD-cellulose for aromatic groups is exploited in order to retard the elution of specific phenoxyacetylated or naphthoxyacetylated aminoacyl-tRNAs. The method generally involves six steps:

1. The crude tRNA is stripped of aminoacyl esters by incubation in Tris buffer at pH 8 to 9 (Sarin and Zamecnik, 1964). The stripped tRNA is then applied to a BD-cellulose column and eluted with 1 *M* NaCl. This step tends to free the crude tRNA from any material that has strong affinity for BD-cellulose.

2. The stripped tRNA is then incubated with the aromatic acylating reagent under the same conditions that are to be used later to react

with aminoacyl-tRNA. The treated tRNA mixture is again applied to a column of BD-cellulose and eluted with 1 *M* NaCl. This step is designed to remove polynucleotides that are susceptible to acylation under the conditions used, and may in some instances be combined with the first step when the material retained on the column is of little interest.

3. The specific tRNA is charged enzymatically with its corresponding amino acid to form a specific aminoacyl-tRNA.

4. The amino group of the aminoacyl-tRNA is reacted with an aromatic acylating reagent such as the phenoxyacetyl ester of N-hydroxysuccinimide or the 2-naphthoxyacetyl ester of *N*-hydroxysuccinimide. This is carried out under conditions that limit aromatic acylation of sites other than the α-amino group of amino acylated tRNA.

5. The tRNA mixture containing the specific aromatic acylated aminoacyl-tRNA is applied to a column of BD-cellulose, and the unaltered tRNA's are eluted with 0.8 *M* NaCl. The substituted aminoacylated tRNA is then eluted with a mixture of 1 *M* NaCl and ethanol.

6. The tRNA is stripped of the substituted aminoacyl derivative by incubation in Tris buffer at pH 8 to 9.

Purified tRNA obtained in this way often consists of more than one species, since any one amino acid may be esterified to several specific tRNAs. Most of these iso-accepting tRNAs are separable by chromatography on BD-cellulose with NaCl gradient elution (Gillam *et al.*, 1967; Seno *et al.*, 1968). However, this still leaves unresolved the question of purity determinations for small quantities of tRNA of high specific radioactivity.*

* One suggestion has been to utilize the ribosomal binding system described by Nirenberg and Leder (1964), and Kaji and Kaji (1964). This (as yet) hypothetical assay would require ribosomes and codon in excess for each tRNA. Depending on the amount of ^{32}P used in the *E. coli* culture, the specific radioactivity of the tRNA may range from 10^5 to 10^6 cpm per μg. With a specific radioactivity of 10^5 cpm per μg, as little as 0.01 μg of tRNA may be assayed. Should the other constituents of the assay system be required in hundredfold excess, this would call for approximately 0.04 μg of nonisotopic codon and 200 μg of nonisotopic ribosomal preparation. The purity of the tRNA would be measured by the percent of the tRNA radioactivity bound by the ribosome-codon system and retained on a millipore filter. Some of the codons for this purpose may be obtained commercially. Others may be prepared from T_1, pancreatic, and U2 RNase digests, or by specific syntheses as described by Uhlenbeck *et al.* (1970).

The following directions are essentially those of Gillam *et al.* (1968). Those portions that have been scaled down to accommodate small quantities of radioactive tRNA are presented in detail.

Materials. Solutions A through F all contain 0.01 *M* $MgCl_2$, 0.01 *M* Na acetate buffer, pH 4.5, and the following additional components:

A	0.3 *M* NaCl
B	0.8 *M* NaCl
C	1.0 *M* NaCl
D	1.0 *M* NaCl and 4.7% ethanol (v/v)
E	1.0 *M* NaCl and 9.5% ethanol (v/v)
F	1.0 *M* NaCl and 19% ethanol (v/v)

Solution A is used to load the tRNA mixtures on the BD-cellulose column. Solutions B and C are primarily used to elute tRNAs that do not contain an aromatic acyl group. Solutions D, E, and F are used to elute aromatic acylated aminoacyl-tRNA.

BD-cellulose may be prepared by the method of Gillam *et al.*, 1967, or obtained from Schwarz BioResearch Inc. If commercial material is used, it should be washed twice in 95% ethanol, twice in 2 *M* NaCl in 25% ethanol, and twice in 2 *M* NaCl or until the $A_{260m\mu}$ of the washing is less than 0.04. The BD-cellulose is then filtered, the filter cake is suspended in 0.4 *M* NaCl, and the slurry is poured into a 0.5 cm × 30 cm glass column. For a column of this size, BD-cellulose of approximately 100 mesh is preferred. The column is washed with 100 ml of 0.3 *M* NaCl or until the $A_{260m\mu}$ of the eluate is below 0.025. The flow rate for this and all other elutions from the BD-cellulose column should not exceed 15 ml per cm^2 per hr, which for a column of 0.5 cm × 30 cm is approximately 12 ml per hr. A number of these small columns may be made simultaneously.

The phenoxyacetyl (and 2-naphthoxyacetyl) ester of N-hydroxysuccinimide is prepared as described by Gillam *et al.* (1968).

Method. 1. The mixture of crude radioactive tRNAs is dissolved in 1 ml of 0.01 *M* Tris buffer, pH 8.5, and incubated for 1 hour at 37°. At the conclusion of the incubation, 0.1 volume of 1 *M* sodium acetate buffer, pH 4.5, is added followed by 2.5 volumes of cold (−20°) 95% ethanol. The chilled suspension is allowed to stand at −20° for

1 hour and then centrifuged at 17,000 × *g* for 30 minutes. The tRNA precipitate is taken up in solution A and applied to a BD-cellulose column. The column is then washed with at least 5 column-volumes of solution A. This is followed by elution with solution C. Most of the free radioactive tRNA should be eluted with this buffer, and elution should continue until the radioactivity reaches an appropriately low value. The fractions containing the radioactive tRNA are combined, dialyzed against water, and then freeze-dried. An alternative method for recovering the tRNA is to dialyze against ethanol and then against water. Dialysis against ethanol reduces the volume of solution in the dialysis sac, the sac is shortened by tying off the end close to the liquid volume, and dialysis is continued against water to remove the ethanol. This procedure may be repeated several times a day, and a decrease in volume as much as 90% may be obtained.

2. The radioactive tRNA from step 1 is dissolved in 1 ml of buffer containing 0.1 *M* triethanolamine-HCl and 0.01 *M* $MgCl_2$, pH 4.3. The solution is cooled to 0° and mixed with 0.1 ml of a 2.4% solution of phenoxyacetylester of N-hydroxysuccinimide in dry tetrahydrofuran. The pH of the mixture is immediately adjusted to 8.0 with 1 *N* NaOH. The correct amount may be determined on a larger volume of buffer and reagent and scaled down accordingly. The mixture is shaken at 0° for 10 minutes and then the pH is adjusted to 4.5 with glacial acetic acid (again the correct amount may be determined on the larger volume). The tRNA is precipitated with 2.5 volumes of chilled (−20°) 95% ethanol and recovered by centrifugation at 17,000 × *g* for 30 minutes. The precipitate is washed twice with cold ethanol, and the ethanol is carefully drained from the pellet.

The pellet of radioactive tRNA is suspended in 1 ml of solution A and applied to a BD-cellulose column. It is eluted and recovered in the same manner as described in step 1.

3. Methods for the preparation of aminoacyl-tRNA synthetases from *E. coli*, yeast, and other sources have appeared in a number of reviews (see articles by Muench and Berg, Calendar and Berg, Baldwin and Berg, Lazzarini and Mehler, Mehler and Stern, Mehler and Jesensky, and Makman in Cantoni and Davies, 1966). While the ideal system for forming a specific aminoacyl-tRNA would contain a single purified synthetase in addition to the specific amino acid and tRNA,

in practice this may not be necessary. A preparation of mixed aminoacyl-tRNA synthetases will do, provided that it is essentially free of any amino acid impurities or amino acid-generating systems.

After the aminoacylation of a specific tRNA has been carried out, the reaction is terminated by adjusting the pH to 5 and cooling the mixture to 4°. Sodium chloride solution is added to yield a final concentration of 0.9 *M* and the aminoacyl-tRNA is precipitated with 2 volumes of 95% ethanol. After standing in the cold, the mixture is centrifuged at 17,000 × *g* for 30 minutes, the supernatant is discarded and the pellet carefully drained to remove as much ethanol as possible.

4. The pellet containing the aminoacylated-tRNA is dissolved in 1 ml of 0.1 *M* triethanolamine buffer and treated with phenoxyacetyl N-hydroxysuccinimide as described in step 2.

5. The tRNA pellet from step 4 containing the phenoxyacetylated aminoacyl-tRNA is suspended in 1 ml of solution A and applied to a BD-cellulose column. The column is washed with solution A and the unesterified tRNA is eluted with solution B. The elution with this solution is continued until the radioactivity of the eluate has reached a low value, usually less than 1% of the peak value. In order to be certain that all of the unesterified tRNA has been removed, the column is washed with several volumes of solution D. The specific phenoxy-acetylated aminoacyl-tRNA is then eluted with solution E. When the 2-naphthoxyacetyl derivative is used as the aromatic moiety, solution F will be needed to elute the specific substituted tRNA from the column because of the stronger binding forces between the naphthoxyacetyl group and the BD-cellulose.

The fractions containing the phenoxyacetylaminoacyl-tRNA are combined and the substituted tRNA is recovered as described in step 1.

6. The substituted tRNA is dissolved in 1 ml of Tris buffer, pH 8.5, and incubated for 1 hour at 37°. At the end of this period, the pH of the reaction mixture is adjusted to 5 with glacial acetic acid, and the tRNA is recovered as before after precipitation with 2.5 volumes of ethanol.

RESOLUTION OF INDIVIDUAL tRNA SPECIES

BD-Cellulose Chromatography. Resolution of the individual tRNA species may be achieved with a column of the same dimensions as that used for the isolation of phenoxyacetylaminoacyl-tRNA. All operations are performed at 4°. A column 0.5 cm × 30 cm is filled with BD-cellulose in a starting buffer of 0.02 *M* sodium acetate, pH 6.0, and 0.5 *M* NaCl. The column is washed with several column-volumes of starting buffer and then the purified tRNA from the previous step is applied to the top of the column. The sample is washed in with several milliliters of starting buffer, and elution is begun with a linear gradient consisting of 50 ml of starting buffer in the mixing vessel, and 50 ml of 0.02 *M* sodium acetate, pH 6.0, and 1.5 *M* NaCl in the reservoir vessel. The flow rate should not exceed 4 ml per hr.

Fractions of 1 ml each are collected and a 0.01-ml portion of each is taken for radioactivity measurement. The peak tubes are combined, dialyzed against water, and then freeze-dried. In many instances, the tRNA at this stage is pure enough to be used in sequence analysis. At present, estimates of purity may be made either from the fingerprint patterns obtained after digestion with specific nucleases or from "dry" runs made with larger quantities of nonradioactive tRNA. In the latter case, the number of moles of amino acid accepted per molar extinction at 260 mμ of the tRNA is used as a measure of purity.

REVERSE-PHASE PARTITION CHROMATOGRAPHY

Somewhat more specialized equipment is required for reverse-phase partition chromatography than is needed for ordinary chromatography. For small-scale applications such as the present one, suitable equipment may be obtained from the Chromatronix Corp. Inc., Berkeley, California. The requirements are for a jacketed, small-bore chromatography column with fittings for elution under pressures as high as 100 psi.

Reverse-phase chromatography differs from ordinary partition chromatography in that in the former the water-insoluble phase is immobilized on the solid support, whereas in the latter the aqueous phase is immobilized on the solid support. The advantages of the former are that a greater range of elution conditions may be utilized by altering the composition of the aqueous eluting solvents. In practice, the water-insoluble quaternary ammonium salts trioctylpropylammonium bromide or dimethyldilaurylammonium chloride are immobilized on a hydrophobic diatomaceous earth support. The sample is applied to the column and eluted with a gradient of sodium chloride. Fractions are collected and portions of each are taken for radioactivity measurement. A number of slightly different systems have been suggested for use in the fractionation of a tRNA mixture (Kelmers *et al.*, 1965; Weiss and Kelmers, 1967; Weiss *et al.*, 1968). However, for the resolution of iso-accepting tRNA species, just two of these are probably sufficient, used either separately or in tandem (Weiss *et al.*, 1968).

Materials. Trioctylpropylammonium bromide is obtained from Eastman Organic Chemicals. Dimethyldilaurylammonium chloride is obtained from General Mills, Chemical Division. Chromosorb-W (acid-washed, dimethyldichlorosilane-treated, 100 to 120 mesh or 200 to 240 mesh) is obtained from Johns-Manville Products Corp.

Preparation of the Column. The directions that follow are essentially the same as given by Weiss *et al.* (1968) except that they have been scaled down for use with small quantities of radioactive tRNA.* They are given for the reverse-phase partition chromatography with trioctylpropylammonium bromide, but the scaling down of the directions with dimethyldilaurylammonium chloride are easily and similarly made. The organic coating solution is made by mixing 0.5 g of trioctylpropylammonium bromide in 50 ml of chloroform. This is mixed with 10 g of Chromosorb-W and stirred in a 200-ml beaker until most of the chloroform is evaporated. The still-damp-with-chloroform mixture is placed in a 100-ml wide-mouthed plastic bottle capped, and tumbled for 2 hours. The cap is removed and the remaining chloroform is allowed to evaporate with continued stirring or mixing. The prepared Chromosorb-W packing is then stored dry until needed. The dry, treated Chromosorb-W is suspended in 20 ml of initial column equilibration solution which consists of 0.15 *M* NaCl, 0.01 *M* buffer, and 0.01 *M* $MgCl_2$. The composition of the buffer is either 0.01 *M* Tris, pH 7.0, or 0.01 *M*

* Experiments by Cliff Singer, University of California, Berkeley, indicate that essentially no radioactive tRNA is lost when the separations are performed on large columns.

sodium acetate, pH 4.5, depending on the iso-accepting tRNAs to be separated.

The suspended Chromosorb-W is poured as a slurry into the microbore column, approximately 2.8 mm × 1000 mm, allowed to settle, and the additions continued until the column is filled with a bed of Chromosorb-W. The Chromosorb-W is then compacted by pressurization, under flow, of about 60 psi. This yields a denser bed for sharper chromatographic peaks. Columns are equilibrated by pumping through two or three column-volumes of initial equilibrium solution. The column is operated at a flow rate up to 5 ml per hr.

Elution is carried out with a linear gradient consisting of 25 ml of 0.3 *M* NaCl, 0.01 *M* $MgCl_2$, and 0.01 *M* buffer (either Tris at pH 7.0 or acetate at pH 4.5) in the mixing vessel and 25 ml of 0.7 *M* NaCl, 0.01 *M* $MgCl_2$, and 0.01 *M* of appropriate buffer. The slope and total volume of this gradient are adjusted, by experiment, to yield the best separation of the iso-accepting species. A preliminary comparison of the separations achieved at two different pHs should be made to determine the best conditions.

Fractions of 0.5 ml are collected and 5-μl portions of them are taken for radioactivity measurement. The purity of the tRNA is determined as described above for the fractions from a BD-cellulose chromatography.

SUMMARY OF THE PREPARATION OF 23S, 16S AND 5S RNA FROM RIBOSOMES

Preparation technique has been adapted from Hosokawa (1970). The ribosomes are separated into 50S and 30S subunits by sucrose gradient centrifugation. The 50S ribosome is extracted with phenol to yield a mixture of 23S and 5S RNA, and these are separated by Sephadex G-100 chromatography. Phenol is used to extract 16S RNA from the 30S ribosome. Each RNA is further purified by gel electrophoresis.

Method. The ribosomal pellet obtained after alumina grinding (and centrifugation) is suspended in 0.9 ml of TMA(−4) buffer, and 0.3-ml aliquots of this suspension are placed on top of each of three tubes containing 4.6 ml of 10 to 30% linear sucrose gradients made in

TMA(−4). The tubes are loaded into an SW-39 rotor and centrifuged at 36,000 rpm for 4 hours at 4°. This step is performed in the same way as that described above for the sucrose gradient centrifugation of tRNA. At the conclusion of the centrifugation, the contents are collected from the bottom of the tubes in 3-drop fractions. A 5-μl portion of each is taken for radioactivity measurement, and the fractions corresponding to the 50S and 30S ribosomes are pooled separately. The material between the two major peaks containing mixed 50S and 30S ribosomes is pooled separately, and sedimented again on a sucrose gradient if desired.

The fractions containing the 50S ribosome are combined and mixed with 0.1 volume of 1 *M* sodium acetate buffer, pH 5.0, 200 μg of bentonite in 0.02 ml, 0.1 volume of a 20% sodium dodecylsulfate, and an equal volume of phenol saturated with 0.1 *M* sodium acetate buffer, pH 5.0. This mixture is shaken or stirred vigorously at room temperature for 15 minutes. Separation of the phases is hastened by centrifugation at 2000 rpm for 10 minutes. The aqueous phase is recovered and extracted with phenol twice more in the same way. This is followed by three extractions with ether and removal of the dissolved ether from the aqueous layer under reduced pressure. The RNA is then precipitated by the addition of 2.5 volumes of 95% ethanol at −20°. The suspension is allowed to stand at −20° for at least 1 hour and then it is centrifuged at 17,000 × *g* for 30 minutes. The RNA pellet is suspended in 1 ml of TMA(−3) buffer and filtered through a Sephadex G-100 column (3.5 cm × 105 cm) with TMA(−3) as eluting buffer. Fractions are collected and portions of each are taken for measurement of the radioactivity.

The first large peak to emerge in the exclusion volume should consist of 23S RNA. A second, smaller peak trailing behind the larger one is 5S RNA. The fractions corresponding to the 23S and 5S RNA peaks are pooled separately, and 100 μg of bentonite are added to each. The solutions may be concentrated with the aid of an Amicon Diaflo apparatus, or by the procedure of alternate dialysis against ethanol and water as described above. This is repeated until the desired reduction in volume is achieved. The RNA is then either recovered from the ethanol as a precipitate or the solution is finally dialyzed against buffer to remove all traces of ethanol.

The phenol extraction of 16S RNA from the 30S ribosome is carried out in the same way as that for the 50S ribosome. Each of these RNA fractions may be purified further by gel electrophoresis.

GEL ELECTROPHORESIS OF rRNA

The first attempts to separate nucleic acids by zone electrophoresis were carried out on starch columns by Paigen (1956). This method was later used by Nomura *et al.* (1960) to identify the messenger RNA "specific to viral infection." For the separation of rRNAs, the variety of solid supports has since expanded to include agar gels (Tsanev, 1965), agarose gels (McIndoe and Munro, 1967), polyacrylamide gels (Loening, 1967; Bishop *et al.*, 1967; Peacock and Dingman, 1967), and a combination of agarose and polyacrylamide gels (Dingman and Peacock, 1968; Peacock and Dingman, 1968). These studies have shown that gel electrophoresis is superior to sucrose gradient centrifugation or methylated albumin-kieselguhr (MAK) column chromatography for the separation of 23S RNA from 16S RNA, from the standpoint of sensitivity, resolution, and time of analysis. As little as 1 μg of nonradioactive RNA is visible, after staining or by UV scanning, in disc electrophoresis; with nonradioactive RNA, 100 μg are needed for sucrose gradient centrifugation or MAK column chromatographic analysis. The sensitivity of the gel electrophoresis may be increased even more through use of capillary tubes or very fine gel threads to permit the analysis of picogram quantities of RNA.

The resolution of the 23S band from the 16S band in gel electrophoresis is often of the order of 6 to 10 bandwidths. In contrast, the separation of these two components is always incomplete by sucrose gradient centrifugation or MAK chromatography; thus additional centrifugations or chromatographic steps are required for analysis and/or purification. In addition, the mobility of an RNA molecule appears to be linearly related to its molecular weight at any one gel concentration (Lewicki and Sinskey, 1970).

The separation of the rRNA components is ordinarily complete after 1 or 2 hours of electrophoresis. The positions of the RNA bands may be determined directly by scanning with a commercial device such as the Joyce-Loebl UV scanner or, in the case of radioactive RNA, the gel is sectioned and the sections are counted. Probably the most time-consuming step is the location of the RNA bands by staining: without the aid of an electric destaining apparatus (such as that manufactured by Canalco, Inc.) this process may run overnight. However, this is balanced by the fact that a dozen or more separate gel electrophoresis runs may be performed simultaneously. Considerably more time and effort would be required to analyze this number of samples by the other methods.

Because of poorer resolution, agar gels have been largely abandoned in favor of agarose and polyacrylamide gels. The separation of 16S and 23S (or 18S and 28S) rRNAs have been performed in polyacrylamide gel concentrations of 2 to 2.6% and in agarose gel concentrations of 2 to 3%. Although the resolution of the 16S RNA from the 23S RNA is better in polyacrylamide gels than in agarose gels, the former are very fluid and difficult to handle. This slight drawback is overcome when the dilute polyacrylamide gel such as a 2% gel is supported by a 0.5% agarose gel (Peacock and Dingman, 1968). Composite gels of this type compare favorably in firmness with those of higher polyacrylamide concentration and are easily sectioned for the location and isolation of radioactive RNA.

The following directions are taken from the study by Peacock and Dingman (1968) and modified for disc electrophoresis.

Materials. Acrylamide and N,N′-methylenebisacrylamide are obtained from Bio·Rad Labs Inc. A purified form of agarose is obtained from Bio·Rad Laboratories or from Bausch and Lomb. Stock solutions are made as follows:

Solution A = buffer consisting of 0.9 *M* Tris, 0.025 *M* EDTA, 0.9 *M* boric acid, pH 8.3
Solution B = acrylamide solution consisting of 19 g acrylamide and 1 g bisacrylamide in 100 ml water
Solution C = 6.0 g dimethylaminopropionitrile (DMAPN) in 100 ml water
Solution D = 1.6 g ammonium persulfate in 100 ml water
Solution E = solution A diluted 1:10 with water

Method. Plexiglas or Lucite tubes are fitted with membranes as described previously for gel electrophoresis of tRNA.

A suspension of 354 mg of agarose in 50 ml of water is refluxed at 100° for 15 minutes with rapid stirring. The solution is allowed to cool to 40° in a water bath. An acrylamide gel solution is prepared from a mixture of 5 ml of solution A, 5 ml of solution B, 3.5 ml of solution C, and warmed to 35°. The acrylamide gel solution is then mixed with 35 ml of the agarose solution and 1.5 ml of solution D. This final composite solution of 2% polyacrylamide and 0.5% agarose is used to fill the Lucite electrophoresis tubes to a height of 11 cm above the bottom. Water is layered carefully over the surface of the gel solution and the tubes are held in a vertical position during the gelation period. Under these conditions, the agarose should solidify first followed by the acrylamide after an hour.

When the gelation is completed, the excess liquid is removed from the top of the gel and replaced with solution E. The tubes are mounted in the reservoirs as described before, and electrophoresis is carried out at 4° (in a cold room) for 30 minutes at a current of about 5 mA per tube. At the end of this period, the current is turned off and the sample is introduced carefully over the surface of the gel. For this purpose, the sample ^{32}P-rRNA is suspended in 10 to 100 μl of 10% sucrose made in solution E and containing 0.02% brom phenol blue. The layering of the sample may be accomplished with a micropipet such as a Lang-Levy pipet fitted with a controlled delivery device. The fine tip of the pipet is introduced over the gel surface and under the buffer, and the contents are slowly expelled to form a narrow layer of dense sucrose solution next to the gel surface. The current is turned on and the electrophoresis is continued at 5 mA per tube until the brom phenol blue marker nears the bottom. The location and extraction of the RNA may be carried out in the same way as described before for tRNA. A parallel electrophoresis of a known mixture of rRNA is often helpful in the identification of the purified 23S or 16S rRNA.

The determination of the purity of the rRNA isolated in the procedure described above possesses many drawbacks. At present, a rough estimate of the homogeneity of the rRNA may be obtained from the appearance of the RNA band in question in a gel or the shape of the rRNA peak in a sucrose gradient centrifugation pattern or an elution pattern from MAK column chromatography. A more meaningful measure of the molecular homogeneity and intactness of the rRNA would be its ability to reconstitute together with ribosomal proteins to form functioning ribosomes. Such a plan leaves a great deal to be desired in terms of simplicity and convenience in an assay. With care in preparation, the gross physical appearance of the nucleic acid may be sufficient assurance of purity. In the final analysis, the clearness with which the results indicate a single sequence is the best measure of purity.

POLYACRYLAMIDE ELECTROPHORESIS OF 5S RNA

The 5S RNA prepared as described above may be further purified by polyacrylamide gel electrophoresis under conditions that are the same as those specified for the electrophoresis of tRNA. The location and isolation of the 5S RNA is carried out in the same manner as that described for tRNA. Some nonradioactive tRNA is included in the electrophoretic run as a marker. The 5S RNA is extracted from the gel as described before for tRNA.

^{32}P-LABELED NUCLEIC ACIDS FROM MAMMALIAN CELLS (RABBIT RETICULOCYTES)

The isolation of ^{32}P-labeled nucleic acid from mammalian sources is considerably simplified when preparations of single cell types are used. At present, single mammalian cell types are obtained from monolayer culture of Hela, KB carcinoma, kidney, and amnion cells derived from humans, and kidney cells from monkeys and cattle. Growth in suspension cultures may be achieved with most of these cell lines (Moore and Ulrich, 1965). Except for 5S RNA (Forget and Weissman, 1968), numerous procedures have been described for the isolation of nucleic acids from mammalian sources. Summaries of these methods have been prepared by Brown (1963), Kirby (1964), Moldave (1967), Moldave and Skogerson (1967), Girard (1967), Kruh (1967), Noll and Stutz (1968), and Penman *et al.* (1969). In the case of mRNA, the messenger for hemoglobin seems to be the most accessible in higher organisms because of the exceptional properties of reticulocytes

(Chantrenne *et al.*, 1967). With rabbit reticulocytes, for example, the hemoglobin mRNAs are long-lived and make up over 85% of the mRNA in the cell. Preparations of blood may be obtained that contain as much as 95% reticulocytes, and these may be freed of other blood cells and the plasma easily (Borsook *et al.*, 1952). They may be lysed in slightly hypotonic solution and are essentially free of ribonuclease.

While the following procedure is centered around the isolation of hemoglobin mRNA, reticulocytes also serve as a source of tRNA, 5S RNA, and rRNA and for this reason have been selected as the example for the isolation of nucleic acids from a mammalian cell. A summary of the procedure is shown schematically in the flow diagram in Fig. 10.4. As before, once the nucleic acid has been freed from the other cellular constituents, the purification steps are the same as those described for nucleic acids from *E. coli.*

Preparation of ^{32}P-labeled Reticulocytes. In order to increase the proportion of reticulocytes in the blood samples, rabbits are injected subcutaneously each day for at least 5 days with 1 ml of a 2.5% phenylhydrazine solution (pH 7.4) (Borsook *et al.*, 1952). Approximately 24 to 48 hours before blood is drawn, the animals are injected intraperitoneally with 5 to 10 mC of ^{32}P-orthophosphate. The blood is collected by heart puncture, chilled, and centrifuged at 3000 rpm for 15 minutes. The chilled cells are washed 3 times in buffer containing 0.13 *M* NaCl, 0.005 *M* KCl, 0.0074 $MgCl_2$, pH 7.4), and each time the thin layer of leucocytes is carefully poured off the surface of the sedimented reticulocytes.

Isolation of ^{32}P-nucleic Acid Fractions. The packed cells from the previous step are suspended in 4.0 volumes of lysis medium (0.01 *M* Tris, 0.015 *M* KCl, 0.005 *M* mercaptoethanol, pH 7.4) containing 0.1 mg of bentonite per ml. After a 90-second incubation, an equal volume of 0.5 *M* sucrose in lysis medium is added and the suspension is centrifuged at 17,000 × *g* for 12 minutes. (This may be repeated until most of the reticulocytes have lysed.) The supernatant is recovered and centrifuged through a cushion of 1 *M* sucrose in lysis medium containing 10 μg of bentonite per ml. The sucrose cushion should occupy between ⅔ and ¾ of the total volume of the centrifuge tube.

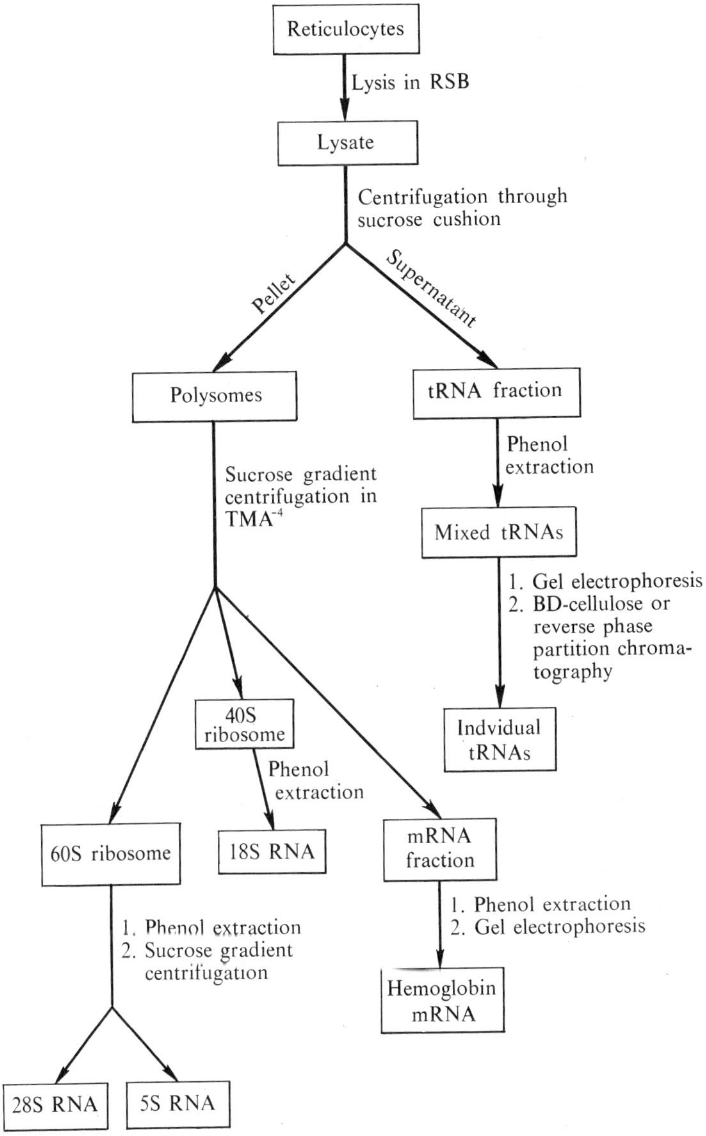

Fig. 10.4. Schematic for the isolation of various RNA species from rabbit reticulocytes.

The centrifugation is performed in a Spinco No. 50 rotor at 48,000 rpm for 3.5 hours at 4° or in a No. 30 rotor at 29,000 rpm for 5 hours.

The supernatant from this centrifugation should contain most of the tRNA fraction with a very small amount of ribosomal material. The tRNA is extracted from the sucrose and purified with phenol as described earlier.

The pellet consists mainly of the polyribosomes (Rich, 1967). It is suspended in TMA(−2) buffer dialyzed for 3 hours against TMA(−4) buffer and layered on a 12 to 30% sucrose gradient in TMA(−4) buffer. The solution is centrifuged at 25,000 rpm for about 20 hours in an SW 25.1 rotor at 2 to 4°. At the end of the centrifugation, the contents of the tubes may be collected from the bottom as described earlier or by upward displacement with a 50% sucrose solution. The effluent is monitored at 260 or 254 mμ, and fractions are collected for assay of radioactivity. The kind of pattern that may be obtained from this sedimentation is shown in Fig. 10.5. The faster sedimenting material consists of the heavy and light ribosomes. These tubes may be combined for the isolation of 5S, 18S, and 28S RNA, as described later. The fraction sedimenting with a constant of about 10S contains the hemoglobin mRNA. The tubes comprising the 8 to 10S peak are combined and mixed with 0.1 volume of 5% SDS containing 0.5 mg of bentonite. The mixture is then extracted at room temperature with an equal volume of phenol saturated with 0.1 *M* sodium acetate buffer, pH 5.0. The RNA in the aqueous phase (on the bottom, owing to sucrose density) is extracted once with ether to remove excess phenol, and it is precipitated with 2.5 volumes of cold 95% ethanol. The mixture is held at −20° for at least 1 hour. The ^{32}P-RNA precipitate is recovered by centrifugation at 17,000 × *g* for 30 minutes and the RNA pellet is taken up in the sample buffer for gel electrophoresis (see Blobel, 1971, for modification using puromycin to dissociate mRNA from the ribosomes).

POLYACRYLAMIDE GEL ELECTROPHORESIS OF HEMOGLOBIN mRNA

From the data of Labrie (1969), it appears that the polyacrylamide gel concentration best suited for the purification of hemoglobin mRNA is between 3 and 4%. At this low gel concentration, the

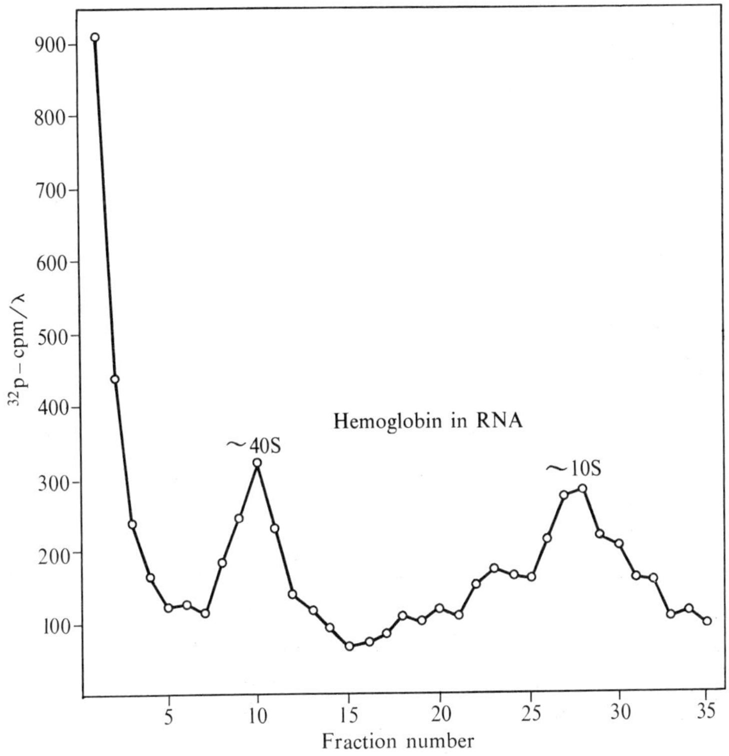

Fig. 10.5. Sucrose gradient centrifugation pattern from reticulocyte polysomes. Centrifugation carried out in 25.3 rotor, 24,000 rpm, 16.5 hours, 4°, in a 12 to 30% sucrose gradient.

advantages of the discontinuous buffer system may realized through use of a dilute agarose gel plug or a dense sucrose solution for the spacer gel. The following directions are for a 3.3% running gel with a 1% agarose spacer gel.

Materials. Buffers A, B, and C are prepared as described for the gel electrophoresis of tRNA. For the 3.3% running gel, the following solutions are prepared:

Solution 1 10.0 g acrylamide
0.6 g bisacrylamide
15 mg potassium ferricyanide
Water to make 100 ml

Solution 2 50 ml buffer A
0.23 ml TEMED
Water to make 100 ml

Solution 3 0.14 g ammonium persulfate
Water to make 100 ml
This solution is degassed with the aid of a water aspirator pump.

Method. The running gel solution is prepared by mixing two parts of solution 1 with one part of solution 2. This mixture is degassed (= solution 4). Mix equal parts of solution 3 and solution 4 and use to load polyacrylamide tubes. After the polyacrylamide running gel has formed, the agarose spacer gel is made by refluxing a 1% suspension of agarose in buffer B that has been diluted 1:8. After refluxing for 15 minutes, the agarose solution is cooled to 45° and then added to the dry surface of the polyacrylamide gel to a height of about 0.5 cm. The tubes are allowed to stand at room temperature until the gelation is complete. The mRNA is taken up in 0.1 ml of sample solution that consists of 10% sucrose, 0.2% brom phenol blue in buffer B diluted 1:8. The sample is applied to the surface of the spacer gel and the electrophoresis is carried out as described before. The mRNA may be located and extracted as described earlier for tRNA.

At present there is some controversy about the sedimentation constant of the hemoglobin mRNA isolated by the various methods used. Labrie (1969) has reported a sedimentation constant of 10S for hemoglobin mRNA isolated essentially as described above, Chantrenne *et al.* (1967) have reported 9S, and Laycock and Hunt (1969a, 1969b) have reported 8S and 10S. These discrepancies may be due in part to the different methods used in the estimation of the sedimentation constant. Ideally, the identification of hemoglobin mRNA should depend on its ability to direct the synthesis of hemoglobin polypeptide in an *in vitro* heterologous system. In view of the fact that this has not been accomplished unequivocally as yet, Labrie (1969) has suggested that a combination of other criteria may suffice temporarily. These criteria include:

1. The kinetics of *in vivo* labeling; mRNA is more rapidly labeled than rRNA.

2. Molecular size; considering that hemoglobin mRNA is probably monocistronic (Slayter *et al.*, 1963; Baglioni, 1963), it should contain at least 438 nucleotides. This indicates a molecular weight in excess of 1.5×10^5.

3. The fingerprint patterns obtained after digestion with specific nucleases should be different from the patterns obtained with known reticulocyte rRNA.

Labrie (1969) has found that the 10S RNA component isolated from polyacrylamide gels had a specific radioactivity fivefold higher than the 5S, 18S, or 29S ribosomal RNA bands. The estimated molecular weight of the 10S RNA is about 1.9×10^5, which is somewhat in excess of the minimum necessary for the β-hemoglobin mRNA chain and accounts for 550 nucleotides. The fingerprint patterns exclude the possibility that the 10S RNA is a degradation product of either the 18S or the 29S RNAs. More extensive sequence analysis may resolve uncertainties concerning the identification of hemoglobin mRNA.

ISOLATION OF 5S, 18S, AND 28S RNA FROM RETICULOCYTE RIBOSOMES

The combined fractions containing the heavy and light ribosomes are dialyzed against several changes of TMA(−2) buffer overnight and then centrifuged at 48,000 rpm in a No. 50 rotor for 4 hours at 4°. The ribosomal pellets are suspended and combined in a minimum amount of TMA(−4) buffer and centrifuged on a 10 to 30% sucrose gradient as described earlier. Fractions are collected and the portions corresponding to the light ribosome are combined and extracted with phenol for the preparation of 18S RNA as described for *E. coli* 16S RNA. The fractions corresponding to the heavier ribosomal particle are combined and extracted with phenol for the preparation of 28S and 5S RNA's. This procedure is also identical with the one for the isolation of 23S and 5S from the *E. coli* 50S ribosome.

PREPARATION OF ^{32}P-TMV RNA; SUMMARY OF THE PROCEDURE

Young tobacco plants are inoculated with TMV, the infected leaves are cut from the plant, and the petioles are immersed in ^{32}P.

The leaves are incubated at room temperature for 6 days and then frozen. The radioactive virus is isolated from the frozen leaves after homogenization and differential centrifugation. The radioactive RNA is recovered by phenol extraction.

Infection of Tobacco Plant with TMV. Young tobacco plants, *Nicotiana tabacum* L., 6 to 12 inches high and containing 4 to 10 leaves, are inoculated with a solution of TMV in 0.001 phosphate buffer, pH 7.5. The concentration of the TMV should be about 100 μg per ml (equivalent to 0.3 $A_{260m\mu}$). The virus solution together with a small amount of carborundum (−400 mesh) may be applied to the upper surface of the leaves with a 1-inch paint brush. The plants are held for about 1 hour, and then the leaves are cut from the plant at the point where the petiole joins the stem. These cuts are made with a razor blade and with a horizontal motion.

The leaves are placed in petri dishes and the cut end of the petiole is immersed in 5 to 10 drops of a ^{32}P solution. Approximately 20 mC of ^{32}P are used for 60 g of leaves. When the ^{32}P solution is absorbed, the cut ends of the petioles are immersed in 5 to 10 drops of tap water and this is continued until the rate of moisture absorption is slowed—about 3 hours. The radioactive leaves are then wrapped in Saran Wrap and placed in a 9 × 13 inch Pyrex baking dish. A second Pyrex baking dish is used as a cover. The leaves are allowed to stand for 6 days at room temperature and under ordinary daily artificial or natural light.

Extraction of ^{32}P-TMV from Leaves. At the conclusion of the incubation, about 60 g of leaves are placed in a plastic bag between two cakes of Dry Ice until the leaves are frozen to the point of brittleness. While still frozen in the bag, the leaves are crushed with the aid of a light mallet or piece of wood until the particles are about 5 mm or less in diameter. The ground leaves are then transferred to a 250-ml Sorvall mixer vessel. Next, 100 ml of a chilled buffer solution is added containing 0.04 *M* EDTA, pH 7.5, and 10 g of $NaHCO_3$. This mixture is blended for about 15 minutes, while the vessel is submerged in a Dry Ice–methoxyethanol bath cooled to between 0 and −10°. The pureed leaves are centrifuged for 30 minutes at 17,000 × *g*. The supernatant is recentrifuged at 17,000 × *g* for 30 minutes, and the supernatant from this is centrifuged at 28,000 rpm for 90 minutes in a No. 30 rotor in a Spinco model L centrifuge. The supernatant is discarded and the pellets are covered with a small amount (2 ml) of 0.002 *M* EDTA-0.02 *M* Na cacodylate buffer, pH 7.5, and allowed to stand overnight to suspend the virus without stirring. The viral suspensions are combined and centrifuged at 17,000 × *g* for 30 minutes to remove particulate impurities, and then at 28,000 rpm in a Spinco No. 30 rotor for 90 minutes to sediment the virus. The viral pellets are taken up in EDTA-cacodylate buffer as before, and the differential centrifugation cycle is repeated except that, before the high-speed step, the viral suspension is filtered through a Nalge membrane filter, 0.45 μ pore size. As much as 20 mg or more of ^{32}P-TMV (specific activity $> 10^6$ cpm/mg) may be isolated from 60 g of infected leaves.

Isolation of ^{32}*P-TMV RNA.* The radioactive RNA is extracted from the TMV by scaling down the quantities used for large-volume preparations. In a typical run, 20 mg of ^{32}P-TMV is suspended in 2 ml of EDTA-cacodylate buffer containing 200 μg of bentonite (prepared by the method of Fraenkel-Conrat *et al.*, 1961). This is mixed with 2 ml of freshly distilled phenol that had been equilibrated with the EDTA-cacodylate buffer, and this mixture is stirred vigorously in a small beaker with a magnetic stirrer for 15 minutes at room temperature. The mixture is centrifuged to break the emulsion, and the upper aqueous layer is reextracted with phenol. The aqueous fraction is then extracted once with ether and the excess ether is removed under vacuum (the low pressure of a water aspirator is sufficient). The radioactive nucleic acid solution is mixed with 0.1 volume of 2 *M* Na acetate buffer, pH 5, and then with 2.5 volumes of 95% ethanol. This is chilled to $-20°$ and is allowed to stand for 2 hours. The solution is centrifuged for 10 minutes at 17,000 × *g* and the pelleted RNA is washed once with 95% ethanol. The RNA pellet is taken up in 0.001 *M* EDTA buffer, pH 7.5, and placed on top of a sucrose gradient 5 to 20% in the same buffer. The gradient is centrifuged in an SW 25 rotor for 10 hours. After centrifugation, the contents are removed through a bottom puncture and collected in fractions of 10 drops. Portions (1 μl of each fraction) are taken for radioactivity measurement. Only those portions of the TMV peak corresponding to the symmetrical gaussian distribution are taken for the ^{32}P-TMV RNA sample.

^{32}P-RNA FROM R17 BACTERIOPHAGE

Subsequent to the discovery of the small RNA-containing bacteriophage f2 (Loeb and Zinder, 1961), a number of other very similar viruses were described, including R17, MS2, fr, and M12. Of this group, the most successful sequence analyses have been carried out with R17 as described in Chapter 8. The preparation of R17 is essentially the same as that of the other viruses of the same group. The following directions for the preparation of ^{32}P-labeled R17 RNA are based on those of Anderson *et al.* (1967), and use the Hfr_1, met^- strain of *E. coli* K-12.

Medium

10 g Difco Bactopeptone
1 g yeast extract
8 g NaCl
Water to make 1 liter; sterilize
Add 2 ml of sterile 1 *M* $CaCl_2$

Two subcultures of *E. coli* K-12, Hfr_1, met^-, are grown in 10 ml each of the above medium. One of these is infected with approximately 10^8 R17 bacteriophage, and the other is used to inoculate the main culture medium. Both the virus-infected subculture and the main medium are incubated at 37°. The main culture is aerated vigorously throughout the incubation. When the OD at 550 mμ of the main culture reaches 0.4 to 0.5 (approximately 4×10^8 bacteria per ml), the virus-infected subculture is added and the vigorous aeration is continued. A small amount of Antifoam B (Dow-Corning) may be added to minimize foaming. The progress of the lysis by R17 may be followed by measuring the decrease in optical density at 550 mμ. When the optical density has reached a minimum (3 to 6 hours after infection), 2 ml of chloroform are added and the aeration is continued for another 10 minutes. Further lysis may be promoted by the addition of 30 ml of 0.2 *M* EDTA, pH 7, 100 mg of egg-white lysozyme and incubation for an additional 30 minutes.

The lysate is cooled and 320 g of ammonium sulfate are added (with stirring) for each liter of cooled lysate. The stirring is continued for 1 hour at 4° and then the suspension is centrifuged at 10,000 × *g* for 30 minutes. Alternatively, the ammonium sulfate of the suspension may be allowed to stand overnight and then centrifuged at 10,000 × *g* for 30 minutes. The precipitate is suspended in 20 ml of SSC buffer (Standard saline-citrate = 0.15 *M* NaCl + 0.015 *M* sodium citrate, pH 7), clarified by centrifugation at 10,000 × *g* for 15 minutes, and then DNase is added at a level of 100 μg per 20 ml of suspension. The suspension is allowed to stand at room temperature for 1 hour, then it is centrifuged at 100,000 × *g* for 90 minutes. The pellets are resuspended in SSC and 6.2 g of CsCl are added for every 10 ml of phage suspension. This suspension is then centrifuged in an SW 39 rotor for 20 to 24 hours at 4°. The suspensions in each tube should be overlaid with mineral oil to minimize evaporation and the danger of the tubes collapsing.

At the end of the centrifuging, the denser, opalescent R17 phage band may be seen below the center of the tube. The tube contents may be collected by puncturing the bottom and collecting drop fractions. When the R17 phage fraction has been collected, the CsCl may be removed by passage of the sample through a column of Sephadex G-50 packed in SSC buffer.

Extraction of ^{32}P-RNA from R17. The RNA is extracted from the purified bacteriophage in the same way that TMV RNA is extracted from TMV. The bacteriophage preparation from above is mixed with 1/10 volume of 20% SDS and 0.1 mg of bentonite for every 100 A_{260} units of bacteriophage. This suspension is then mixed with an equal volume of SSC-saturated phenol (redistilled phenol that has been saturated with SSC). The mixture is stirred vigorously for 15 minutes at room temperature and then centrifuged at low speed to break the emulsion. The upper aqueous layer is aspirated off and extracted two more times with SSC-saturated phenol. The excess phenol is removed from the aqueous layer by two extractions with ether and the ^{32}P-RNA is precipitated by the addition of 2.5 volumes of ethanol. The suspension is allowed to stand at −20° for at least an hour, and then it is centrifuged at 17,000 × *g* for 30 minutes. The clear ethanolic supernatant is discarded and the pellet may be used for sequence analysis immediately or may be stored at −20° in a closed tube for several weeks.

REFERENCES

Anderson, J. S., Dahlberg, J. E., Bretscher, M. S., Revel, M., and Clark, B. F. C. (1967). *Nature*, **216**, 1072.

Bachrach, H. L., and Breese, S. S., Jr. (1968). In *Methods in Virology*, Vol. 4. K. Maramorasch and H. Kaprowski (eds.). Academic Press, New York.

Baglioni, C. (1963). In *Molecular Genetics*. J. H. Taylor (ed.). Academic Press, New York.

Bishop, D. H. L., Claybrook, J. R., and Spiegelman, S. (1967). *J. Mol. Biol.*, **26**, 372.

Blobel, G. (1971). *Proc. Natl. Acad. Sci.*, **68**, 832.

Borsook, H., Deasy, C. L., Haagen-Smit, A. J., Keighley, G., and Lowy, P. H. (1952). *J. Biol. Chem.*, **196**, 669.

Brown, G. L. (1963). In *Progress in Nucleic Acid Research*, Volume 2. J. N. Davidson and W. E. Cohn (eds.). Academic Press, New York.

Cantoni, G. L., and Davies, D. R. (eds.) (1966). *Procedures in Nucleic Acid Research*. Harper and Row, New York.

Chantrenne, H., Burny, A., and Marbaix, G. (1967). *Prog. Nucl. Acid Res. Mol. Biol.*, **7**, 173.

Chargaff, E., and Davidson, J. N. (eds.), Volumes 1 and 2 (1955); Volume 3 (1960). *The Nucleic Acids*, Academic Press, New York.

Cherayil, J. D., and Bock, R. M. (1965). *Biochemistry*, **4**, 1174.

Colowick, S. P., and Kaplan, N. O. (eds.) (1960). *Methods in Enzymology*, Vol. 5. Academic Press, New York.

Davidson, J. N., and Cohn, W. E. (eds.), Vols. 1–10 (1963–1970). *Prog. Nucl. Acid Res. Mol. Biol.* Academic Press, New York.

Dingman, C. W., and Peacock, A. C. (1968). *Biochemistry*, **7**, 659.

Forget, B. G., and Weissman, S. M. (1968). *J. Biol. Chem.*, **243**, 5709.

Fraenkel-Conrat, H., Singer, B., and Tsugita, A. (1961). *Virology*, **14**, 54.

Garen, A., and Levinthal, C. (1960). *Biochim. Biophys. Acta*, **38**, 470.

Gillam, I., Millward, S., Blew, D., von Tigerstrom, M., Wimmer, E., and Tener, G. M. (1967). *Biochemistry*, **6**, 3043.

Gillam, I., Blew, D., Warrington, R. C., von Tigerstrom, M., and Tener, G. M. (1968). *Biochemistry*, **7**, 3459.

Girard, M. (1967). In *Methods in Enzymology*. L. Grossman and K. Moldave (eds.), Vol. 12-A. Academic Press, New York.

Grossman, L., and Moldave, K. (eds.) (1967 and 1968). *Methods in Enzymology*, Vols. 12-A and 12-B. Academic Press, New York.

Holper, J. C. (1969). In *Fundamental Techniques in Virology*. K. Habel and N. P. Salzman (eds.). Academic Press, New York.

Hosokawa, K. (1971). *J. Biol. Chem.*, in press.

Kaji, H., and Kaji, A. (1964). *Proc. Natl. Acad. Sci.*, **52**, 1541.

Kawade, Y., Okamoto, T., and Yamamoto, Y. (1963). *Biochem. Biophys. Res. Commun.*, **10**, 200.

Kelmers, A. D. (1966). *J. Biol. Chem.*, **241**, 3540.

Kelmers, A. D., Novelli, G. D., and Stulberg, M. P. (1965). *J. Biol. Chem.*, **240**, 3979.
Kirby, K. S. (1964). In *Prog. Nucl. Acid Res. Mol. Biol.* J. N. Davidson and W. E. Cohn (eds.), Vol. 3. Academic Press, New York.
Kruh, J. (1967). In *Methods in Enzymology*. L. Grossman and K. Moldave (eds.), Vol. 12-A. Academic Press, New York.
Labrie, F. (1969). *Nature*, **221**, 1217.
Laycock, D. G., and Hunt, J. A. (1969a). *Nature*, **221**, 1118.
Laycock, D. G., and Hunt, J. A. (1969b). *Cold Spring Harbor Symp. Quant. Biol.*, **34**, 579.
Lewicki, P. P., and Sinskey, A. J. (1970). *Anal. Biochem.*, **33**, 273.
Loeb, T., and Zinder, N. D. (1961). *Proc. Natl. Acad. Sci.*, **47**, 282.
Loening, H. E. (1967). *Biochem. J.*, **102**, 251.
McIndoe, W., and Munro, H. N. (1967). *Biochim. Biophys. Acta*, **134**, 458.
Miyazaki, M., and Takemura, S. (1966). *J. Biochem.*, **60**, 526.
Moldave, K. (1967). In *Methods in Enzymology*. L. Grossman and K. Moldave (eds.), Vol. 12-A. Academic Press, New York.
Moldave, K., and Skogerson, L. (1967). In *Methods in Enzymology*. K. Grossman and K. Moldave (eds.), Vol. 12-A. Academic Press, New York.
Moore, G. E., and Ulrich, K. (1965). *J. Surg. Res.*, **5**, 270.
Muench, K. H., and Berg, P. (1966). *Biochemistry*, **5**, 970.
Nirenberg, M., and Leder, P. (1964). *Science*, **145**, 1399.
Nishimura, S., Harada, F., Narushima, V., and Seno, T. (1967). *Biochim. Biophys. Acta*, **142**, 133.
Noll, H., and Stutz, E. (1968). In *Methods in Enzymology*. L. Grossman and K. Moldave (eds.), Vol. 12-B. Academic Press, New York.
Nomura, M., Hall, B. D., and Spiegelman, S. (1960). *J. Mol. Biol.*, **2**, 306.
Paigen, K. (1956). *Anal. Biochem.*, **28**, 284.
Peacock, A. C., and Dingman, C. W. (1967). *Biochemistry*, **6**, 1818.
Peacock, A. C., and Dingman, C. W. (1968). *Biochemistry*, **7**, 668.
Penman, S., Greenberg, H., and Williams, M. (1969). In *Fundamental Techniques in Virology*. K. Habel and N. P. Salzman (eds.), Academic Press, New York.
Pestka, S. (1968). *J. Biol. Chem.*, **243**, 4038.
Porterfield, J. S. (1967). In *Methods in Virology*, Vol. 1. K. Maramorasch and H. Kaprowski (eds.). Academic Press, New York.
Rich, A. (1967). In *Methods in Enzymology*. L. Grossman and K. Moldave (eds.), Vol. 12-A. Academic Press, New York.
Richards, E. G., Coll, J. A., and Gratzer, W. B. (1965). *Anal. Biochem.*, **12**, 452.
Sanger, F., Brownlee, G. G., and Barrell, B. G. (1965). *J. Mol. Biol.*, **13**, 373.
Sarin, P. S., and Zamecnik, P. C. (1964). *Biochim. Biophys. Acta*, **91**, 653.
Seno, T., Kobayashi, M., and Nishimura, S. (1968). *Biochim. Biophys. Acta*, **169**, 80.
Slayter, H. S., Warner, J. R., Rich, A., and Hall, C. E. (1963). *J. Mol. Biol.*, **7**, 652.
Tanaka, K., Richards, H. H., and Cantoni, G. L. (1962). *Biochim. Biophys. Acta*, **61**, 846.
Tsanev, R. (1965). *Biochim. Biophys. Acta*, **103**, 374.
Uhlenbeck, O. C., Baller, J., and Doty, P. (1970). *Nature*, **225**, 508.
Weiss, J. F., and Kelmers, A. D. (1967). *Biochemistry*, **6**, 2507.
Weiss, J. F., Pearson, R. L., and Kelmers, A. D. (1968). *Biochemistry*, **7**, 3479.
Yaniv, M., and Barrell, B. G. (1969). *Nature*, **222**, 278.

11.

BASIC SEQUENCE ANALYSIS PROCEDURES

THE OVERLAP METHOD

At present, the greatest proportion of sequence determinations have been made with the overlap method. Given a sample of radioactive nucleic acid such as ^{32}P-tRNA, the first phase of the application of this method is to assemble catalogs of oligonucleotides produced by specific nuclease action. The assembly of the catalogs is ordinarily carried out in the following steps:

1. Digestion of the nucleic acid with specific nuclease.
2. Separation of the oligonucleotides in the digestion mixture.
3. Identification and determination of the oligonucleotides.

The first step is straightforward and the conditions have been well established through widespread use. For the second and third steps, early sequence work was carried out with some modification of the column chromatographic procedures of Volkin and Cohn (1953), and Tomlinson and Tener (1963), and the two-dimensional paper chromatography of Rushizky and Knight (1960a,b). (See also Rushizky *et al.*, 1961; Rushizky and Sober, 1962.) More recently, the two-dimensional fingerprint method of Sanger *et al.* (1965) has provided the most convenient method for assembling oligonucleotide catalogs.

As described briefly in Chapter 4, the oligonucleotides in a nuclease digestion mixture are first separated by ionophoresis on a strip of cellulose acetate. Next, the material on the strip is transferred to a sheet of DEAE-paper and the oligonucleotides are then further separated by ionophoresis at low pH. This second separation is carried out at a right angle to the original direction. An autoradiograph is made of the DEAE-paper sheet to show the positions of the radioactive spots. With this method, separations may be obtained of *sequence* isomers up to a chain length of 4 or *more*.

The individual spots are eluted and, provided no additional purification step is indicated, identified and the molar concentration with respect to the total radioactivity determined. Typical fingerprints obtained with T_1 and pancreatic RNase digests of STNV RNA are shown in Figs. 11.1 and 11.2. The similarities in the patterns from independent laboratories demonstrate that identifications of monomers, dimers, and trimers may be made solely by the positions occupied in the fingerprint.

Size of Sample. In order to achieve good resolution in the separation patterns, the amount of nucleic acid used for each determination should not exceed 100 μg. The total radioactivity of the sample that is required depends on the size of the nucleic acid. Pancreatic and T_1 RNase digests of tRNA or 5S RNA may produce 20 to 40 different spots. A level of radioactivity of 2×10^5 to 2×10^6 per sample is probably sufficient in this instance. With higher molecular weight RNAs such as rRNA, it is expected that more than 100 different spots may need to be identified and determined (Fellner *et al.*, 1970; Fellner and Ebel, 1970). Consequently, material of higher specific activity is required to compensate for the additional time involved in the determinations and the greater variety of oligonucleotides. A level of specific activity of 1000 cpm per nucleotide residue is the minimum.

Digestion Conditions. Both pancreatic and T_1 RNases may be obtained in highly purified forms from commercial sources. Pancreatic RNase from Worthington (RAF), Calbiochem (55674), or Sigma (XII-A) is essentially devoid of extraneous nucleases and no advantage is to be gained by further purification. This is true also for T_1 RNase as supplied by Calbiochem (556785) or Worthington (RT). The T_1 RNase from Sigma, while devoid of extraneous nuclease, has contained

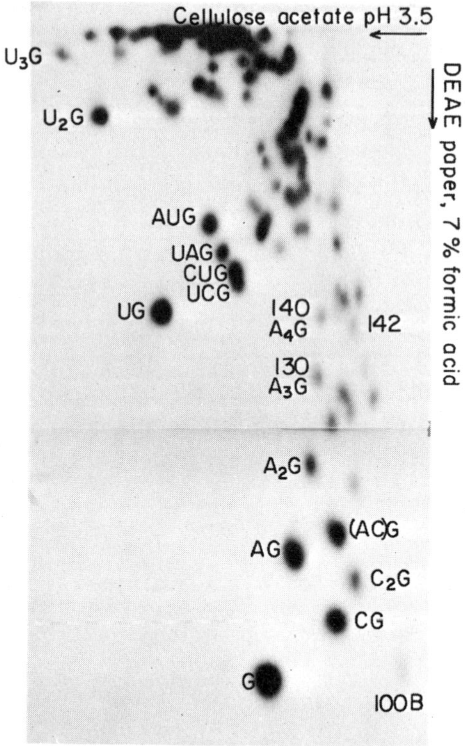

Fig. 11.1. Fingerprint of T_1 RNase digest of satellite tobacco necrosis virus RNA.

phosphomonoesterase activity. This may be removed after a single phenol extraction step (Rushizky *et al.*, 1963) or by performing the hydrolyses in the presence of 0.1 *M* phosphate buffer to inhibit the phosphomonoesterase activity.

Even with very pure nuclease preparations, nonspecific nucleolytic breaks may occur after prolonged periods of hydrolysis or in the presence of excess enzyme. On the other hand, if the digestion period is too short or if too little enzyme is used, many oligonucleotides will terminate in a 2′,3′-cyclic phosphate intermediate product, instead of the final 3′-phosphate. These effects may be balanced by keeping within somewhat narrow limits of enzyme digestion conditions.

Finally, for complete hydrolysis of all susceptible bonds, the extent of secondary structure may be minimized by including a chelating agent such as EDTA in the reaction mixture and by performing the digestion at as high temperatures as is feasible. Ordinarily, temperatures of 40° or 45° are acceptable considering the thermal stability of pancreatic and T_1 RNases.

A suitable buffer for pancreatic or T_1 RNase hydrolysis of nucleic acid is 0.01 *M* Tris, 0.001 *M* EDTA, pH 7.4. The enzyme-to-substrate ratio should be about 50 enzyme units per mg of RNA or 2 enzyme units per OD unit ($A_{260m\mu}$-cm^2) of RNA. The time of incubation at 37°

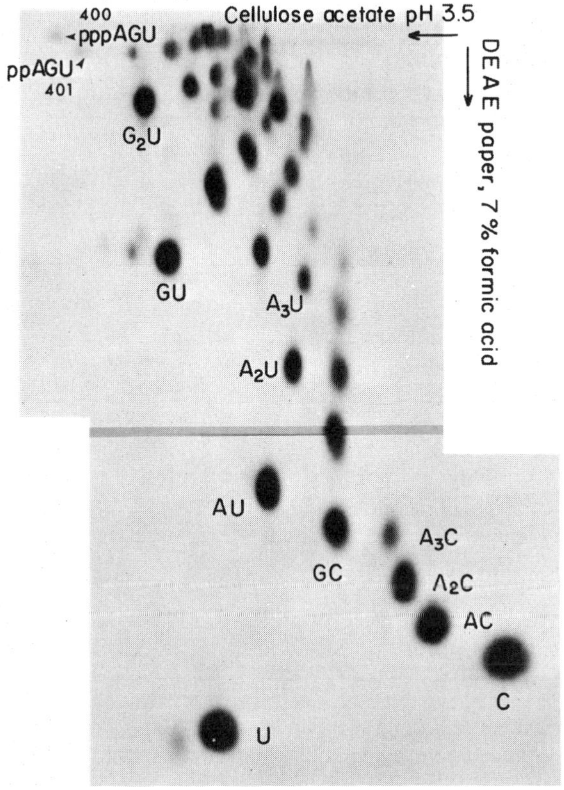

Fig. 11.2. Fingerprint of pancreatic RNase digest of satellite tobacco necrosis virus RNA.

may be as short as 30 minutes (Sanger *et al.*, 1965), but may still be as long as 4 hours without undue nonspecific bond cleavage (Lloyd and Mandeles, 1970). The concentration of nucleic acid may range from 1 μg per μl to 10 μg per μl.

Nuclease digestions may be carried out in a capillary melting point tube (approximately 2 mm × 100 mm) with one end drawn out to a point suitable for spotting. During the incubation, the capillary tube is placed with its pointed tip down in a test tube that has been treated with silicone.*

If the volume of the reaction mixture is 20 μl or less, it may be spotted onto the cellulose acetate strip with no further treatment. For larger volumes, the mixture may be freeze-dried and then diluted with water to 20 μl or less for spotting. The presence of the enzyme does not seem to alter the mobility of the oligonucleotides appreciably. In instances where the enzyme must be removed, bentonite may be added to the reaction mixture in amounts of 1 μg of bentonite per unit of enzyme. This serves to complex both pancreatic and T_1 RNase so that they will not move from the origin of the cellulose acetate or the DEAE-paper.

In the separation of the T_1 RNase oligonucleotides by this fingerprint procedure, the base residue that exerts the greater effect on the mobility in the DEAE-paper dimension is U. The fingerprint appears to be divided into sections containing one, two, three, four, or more U residues. This is shown in a diagram in Fig. 11.1. Fractionation of oligonucleotides that contain two or more U residues is poor because their mobility in the DEAE-paper dimension is less. In order to increase the mobility of these oligonucleotides, an enzymatic dephosphorylation with PME may be carried out at the same time as the T_1 RNase digestion. The dephosphorylated oligonucleotides tend to move to the positions occupied by the phosphorylated oligonucleotides that contain one less U residue—in a sense, moving all of the nucleotides "down." The simultaneous dephosphorylation and nuclease digestion is carried

* Most of the small test tubes and vessels that are involved in the preparation of the small amounts of radioactive RNA should be siliconized. This is accomplished by filling the clean tubes with a 1% solution of dimethyldichlorosilane in benzene (available from Bio·Rad Laboratories). The solution is then returned to the reagent bottle and the tubes are rinsed with distilled water. This is followed by drying in an oven.

out under the same conditions as the digestion with T_1 RNase alone. The PME (BAPF as supplied by Worthington Biochemical Corp.) is used in a quantity corresponding to 10 enzyme units per $OD_{260m\mu}$ per unit of RNA.

U2 RNase

A ribonuclease, U2 RNase, with a specificity for purine nucleotides analogous to the specificity of pancreatic RNase for pyrimidine nucleotides may be isolated from the smut fungus *Ustilago sphaerogena* (Arima *et al.*, 1968a,b). Because the commercial production of this enzyme is imminent, directions for its preparation are not given here. This enzyme is especially useful for the sequence analysis of oligonucleotides from T_1 RNase digests, behaving in this instance as an A-specific enzyme. It has been used by Adams *et al.* (1969) in the sequence analysis of a fragment of coat protein cistron from R17 RNA, and by Gupta *et al.* (1970) for a portion of the coat protein cistron of F2 RNA.

The conditions suggested for the complete hydrolysis of nucleic acid to oligonucleotides terminated with a purine are those of Uchida *et al.* (1969):

A solution containing 50 μg to 1 mg of RNA in 0.2 ml of 0.05 *M* acetate buffer, pH 4.5, is digested for 16 hours at 37° with 0.5 unit of U2 RNase per mg of RNA. The extent of cleavage adjacent to A and G is virtually complete and practically nil for U and C.

TWO-DIMENSIONAL ELECTROPHORESIS

Materials. Rolls of Celotate brand of cellulose acetate in widths of 2 inches and lengths of 10 feet may be obtained from the Millipore Filter Company. Other companies such as Oxoid in England supply it as sheets that must then be cut (with difficulty, since cellulose acetate is very brittle when dry) into strips of appropriate length.

DEAE-paper is obtained as Whatman DE-81 from Reeve-Angel Company. This grade of DEAE-paper is difficult to handle when wet. As an alternative, this paper may be obtained with a plastic backing. In our laboratory, no decrease in resolution is noticed with this product with a considerable improvement in its wet strength. These products are available in rolls 18 inches wide and over 150 feet long.

At least two sets of electrophoresis tanks and racks should be obtained. One should be large enough to accommodate a sheet of paper 10 inches wide and 35 inches long; this may be used for running up to four strips of cellulose acetate simultaneously. The second should be able to accommodate a sheet of paper 20 inches wide and 35 inches long for use with the DEAE-paper. Electrophoresis equipment of suitable dimensions may be obtained from Savant Company. The power supply should be capable of an output of at least 3000 volts at 200 mA. This is also supplied by Savant.

"No Screen" X-ray film is obtained from Eastman Kodak Company in 14 × 17 inch sizes. This film is two to three times more sensitive than Kodak Blue Brand. X-ray exposure cassettes to accommodate this size are also obtained from Eastman Kodak. Varsol, a coolant, is obtained from Savant Instruments, Inc. Marker dyes consist of a mixture of Xylene cyanol F.F. (blue), acid fuchsin (red), and methyl orange (yellow), available from J. T. Baker Chemical Co.

The buffer for the cellulose acetate dimension consists of 0.5% pyridine-5% acetic acid-0.001 *M* EDTA in 7 *M* urea, pH 3.5. The buffer for the DEAE-paper dimension is 7% formic acid, pH 1.9, although a mixture of 8.7% acetic acid and 2.5% formic acid is effective for the smaller oligonucleotides. Radioactive ink for marking autoradiograms may be made of a mixture of ordinary ink and a nonvolatile ^{14}C solute.

All of the following directions are based on the use of the Savant or other equipment of similar dimensions.

Method. The electrophoresis tank reservoirs are filled with buffer (pyridine-acetic acid-EDTA-7 *M* urea for cellulose acetate, or 7% formic acid for DEAE-paper) and the remainder of the tank is filled with Varsol. Water is circulated through the cooling coils to help dissipate the heat evolved at the surface of the cellulose acetate or DEAE-paper during the electrophoresis. The cellulose acetate is cut to a length that is sufficient to go over the rack and extend into the buffer in the reservoirs at each end. For some tanks this length is either 24 or 36 inches. Next, the cellulose acetate strip is marked on one edge to indicate the position of the sample. This is done lightly with a wax pencil about 2½ inches from one end of the strip. This assures that the sample spot will clear the upper surface of the buffer in the reservoir when the rack is placed in position. The cellulose acetate strip is immersed in buffer for a few minutes, then dipped into Varsol, and secured on the rack. The rack is placed in the tank and the power is turned on for a preliminary electrophoresis for 30 minutes at 2000 volts. After this, the rack is removed from the tank, placed on its side, and excess Varsol blotted away from the sample area. The sample is then applied to the center of the strip, on a level with the premarked spot, with the aid of a micropipet or other capillary with a fine tip. A spot of indicator dye is placed on either side of the sample spot and the rack is returned to the electrophoresis tank. The electrophoresis is allowed to continue until the red dye marker has moved 16 to 18 inches from the origin. At 2000 volts, the red dye moves at about 2½ inches per hour. At the end of the electrophoresis, the rack is removed from the tank and the excess Varsol is allowed to drain off. While still damp with buffer, the material on the cellulose acetate strip is transferred to DEAE-paper. For this purpose, a sheet of DEAE-paper is cut from the roll to a length of about 35 inches and marked in pencil as shown in Fig. 11.3, with the right-hand margin marked off at 1-inch intervals for about 12 inches. The paper is placed on a piece of plate glass, and the damp cellulose acetate strip is placed on top. The lower border of the cellulose acetate strip should be positioned just above the transverse margin of the DEAE-paper. Three strips of 3 MM paper 2 inches wide and 18 inches long are prepared. They are soaked in water, the excess moisture removed, and the strips placed on top of the cellulose acetate strip. On top of this is placed a sheet of methacrylic plastic (Plexiglas) measuring 1 × 5 × 25 inches. Pressure is applied by hand to the plastic weight to force the liquid out of the 3 MM paper through the cellulose acetate strip and into the DEAE-paper. As the water passes through the cellulose acetate, it elutes the radioactive oligonucleotides into the DEAE-paper. This transfer is usually accomplished without appreciable spreading because of the ion-exchange binding property of the DEAE-paper. The plastic weight is removed, the 3 MM strips are wetted again with water from a squeeze bottle, the plastic weight is replaced, and pressure is applied as before. Over 90% of the material on the cellulose acetate may be transferred by these means.

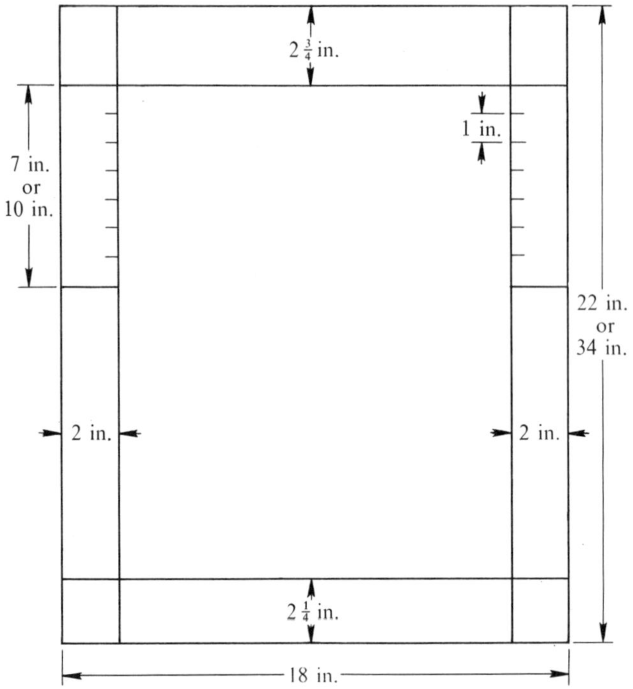

Fig. 11.3. Penciled guide lines for DEAE-paper dimensions.

The 3 MM strips and the cellulose acetate are then gently removed and wet DEAE-paper is carefully lifted off the surface of the glass plate. This is accomplished by grasping the dry end and peeling the wet portion off the surface of the glass. The wet portion of the DEAE-paper is placed in a plastic tray that contains about 800 ml of 95% ethanol and is large enough to accommodate the width of the sheet. After a few moments of contact with the ethanol (actually washing), the sheet is transferred to a second tray containing fresh ethanol and washed again. This washing may be repeated a third time, after which the sheet is mounted on an electrophoresis rack and dried in air. The ethanol serves to remove excess urea and also to remove the water from the wet portions so that the paper can be manipulated more readily.

Wetting of the DEAE-paper on the rack with 7% formic acid solution is accomplished in the following way: except for the area that contains the transferred sample, most of the paper is wetted by directing a stream of 7% formic acid from a squeeze bottle over the surface. This may also be performed with a spraying apparatus. Finally, the sample area is wetted by directing a gentle spray over the surface until the paper is wet. A spot of dye is placed on the outer margins marked in pencil and in line with the sample. The rack is then placed gently into the electrophoresis tank and the power is turned on. Electrophoresis may be carried out at voltages as high as 1500 volts and currents as high as 120 mA, until the blue marker dye has moved 0.33 of the total distance (for pancreatic RNase digests) or 0.4 of the total distance (for T_1 RNase digests). For a sheet of paper 36 inches long and 18 inches wide, the product of the voltage and current should not exceed 200 watts. With a total path of 33 to 35 inches, and a potential of 700 volts, the blue marker will move about 10 inches in 24 hours.

At the end of the electrophoresis, the power is turned off and the rack is lifted gently out of the tank and the paper allowed to dry in place. The positions of the blue markers and sample numbers are noted with radioactive ink, and the paper is cut into two pieces for autoradiography. In general, the time of exposure may be as short as 8 hours when 3×10^5 cpm of ^{32}P are spotted on the cellulose acetate strip. The positions of the radioactive spots on the DEAE-paper are outlined with the aid of the exposed film. The spots are numbered for identification and cut out.

When the identity of the oligonucleotide in a spot is known, the amount of radioactivity may be determined directly by immersing the paper spot in a vial of scintillation fluid and counting it. For determining the identity of an oligonucleotide in a spot, elution may be carried out after counting since the scintillation fluid will not dissolve the hydrophilic oligonucleotide. The elution may be accomplished with a solution of 2 *M* TEA-HCO_3 containing a mixture of 100 μg each of adenosine-3′(2′)-phosphate, cytidine-3′(2′)-phosphate, guanosine-3′(2′) phosphate, and uridine-3′(2′)-phosphate per milliliter of TEA-HCO_3.

Elution of Radioactive Spots from DEAE-Paper. The method suggested by Sanger *et al.* (1965) for elution of oligonucleotides from DEAE-paper is based on the simple device shown in Fig. 11.4 (Sanger and Tuppy, 1951). The elution solvent used is approximately 3.7 *M* TEA-HCO_3, and it is made by passing CO_2 through a 3.7 *M* solution

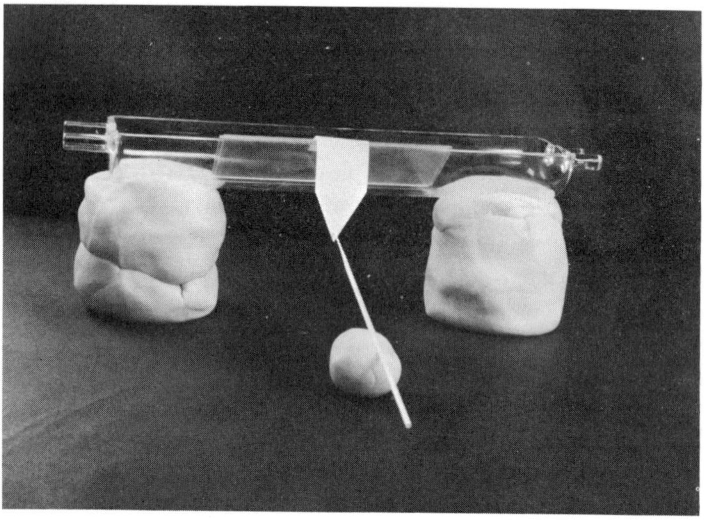

Fig. 11.4. Device used in elution of oligonucleotides from DEAE-paper.

of freshly distilled TEA until the pH drops to about 9. During elution of oligonucleotides from the DEAE-paper, most of the radioactive material is present in the first 20 μl. As a consequence, it is not necessary to elute with more than 50 to 100 μl of solvent. After elution, the sample is dried in a polyethylene planchet under partial vacuum (in a desiccator). When desired, the dried sample may be counted in a gas flow counter. Additional analytical procedures may then be performed such as digestion with alkali, or endo- and exonucleases.

FRACTIONATION OF LONGER OLIGONUCLEOTIDES

While the fingerprint method described above is very effective for the fractionation of oligonucleotides up to a chain length of 4 to 5, the resolution of longer oligonucleotides of similar base compositions is noticeably poorer. As an answer to this problem, Brownlee *et al.* (1968) devised the technique of homochromatography. This is also a two-dimensional fingerprint method that utilizes ionophoresis on cellulose acetate for the first dimension and displacement chromatography for the second dimension. A solution of unlabeled nucleotides is used as the displacing solvent in the second dimension which was carried out on DEAE-paper. Later, Brownlee and Sanger (1969) modified the procedure so that the second dimension fractionation was performed on thin-layer DEAE-cellulose plates rather than DEAE-paper. This improved the resolution of the oligonucleotides so that fragments containing as many as 50 residues could be fractionated. The following is essentially the procedure as given by Brownlee and Sanger (1969).

Materials. DEAE-cellulose (MN 300) and cellulose (MN 300) are obtained from Brinkmann Instruments or Analtech Corp. These celluloses are used in two kinds of mixtures: one mixture contains 1 part of DEAE-cellulose to 10 parts of cellulose, and the other contains one part of DEAE-cellulose to 7.5 parts of cellulose. The 1:10 mixture is used for the fractionation of longer oligonucleotides that do not separate well on the 1:7.5 mixture. Because reasonably priced plates are available in any thickness and any DEAE-cellulose:cellulose composition from the Analtech Corp., no further description of their preparation will be made here.

Solvents for the chromatographic development of the thin-layer plates by displacement chromatography consist of nucleic acids or oligonucleotides dissolved in 7 *M* urea. Three mixtures are used:

Mixture *a* is prepared by dissolving 10 g of commercial yeast RNA in 200 ml of 7 *M* urea and adjusting the pH to 7.5.

Mixture *b* is prepared by dialyzing mixture *a* against 7 *M* urea for 2 to 3 hours at 4°.

Mixture *c* is prepared by dissolving 10 g of yeast RNA in 100 ml of 1 *N* KOH and allowing it to stand at room temperature for 15 minutes. This solution is then neutralized to pH 7.5 with concentrated HCl and dialyzed against distilled water for 2 to 4 hours. Urea (84 g) is added and the volume of the solution is adjusted to 200 ml to yield a 5% solution of oligonucleotides in 7 *M* urea. Brownlee and Sanger (1969) point out that this solution may be diluted with 7 *M* urea to give a 3% solution of oligonucleotides. This provides sharper spots, although the R_f values are lower. Mixture *c* is useful in the separation of oligonucleotides up to a chain length of about 20, while mixtures *a* and *b* are useful for chain lengths of 15 to 50 (mixture *b* apparently provides better resolution than *a*, possibly owing to the lower salt content).

Another generally useful solvent, similar to *b*, is prepared by dissolving 50 g of yeast RNA in 500 ml of water. The yeast is added gradually to the water with constant stirring, and the pH is maintained between 7 and 8 by the addition of 10 *N* KOH. When the solution is complete, the pH is adjusted to 7 and the solution is dialyzed against water in a continuous counterflow dialyzer (single-coil unit, Omnivector Instruments). The time of dialysis is 2 hours for every 200 ml of solution. At the conclusion of the dialysis, the RNA solution is collected and reduced in volume to 500 ml, 420 g of reagent urea are added, and the volume is adjusted to 1 liter with water. This modification has the advantage of being adaptable to the large scale, reproducible preparation of homochromatography solvent.

Method. In carrying out the fingerprint procedure with thin-layer homochromatography, the first dimension separation on cellulose acetate is accomplished in the same way as before. The long oligonucleotides are usually located in the region after the blue marker and before the red marker, and it is this portion of the strip that is transferred to the thin-layer DEAE-cellulose plates. Plates measuring 20 × 40 cm are used. The still damp portion of the cellulose acetate strip to be transferred is placed gently across the thin-layer plate, about 2 cm from one edge (and perpendicular to the direction of the homochromatographic development). This is carefully overlaid with three moist strips of Whatman 3 MM paper. A glass or plastic plate is placed on top, and the "sandwich" is allowed to stand for 15 minutes. At the end of this time, the weighted glass or plastic, the strips of Whatman 3 MM paper, and the cellulose acetate are carefully removed, and the thin-layer plate is allowed to dry in air. The plate is then rinsed in absolute ethanol to remove excess urea and allowed to dry again.

The homochromatography is carried out at 60°. For this purpose, both the thin-layer plate and the chromatographic tank containing the solvent are preincubated at 60° for about 30 minutes. The plate is then placed inside the tank, the tank is covered, and the development proceeds by ascending chromatography. When the front reaches the top (plates 100 μ thick require 2 to 4 hours for this, plates 250 μ thick

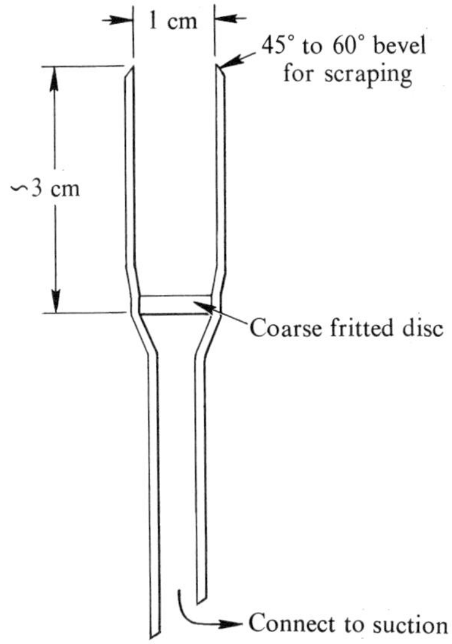

Fig. 11.5. TLC scraper modified from small fritted disc funnel.

require as long as 9 hours), the plate is removed, thoroughly dried, and placed in a cassette with X-ray film as in the fingerprint procedure with DEAE-paper.

The spots to be eluted are located with the aid of a template made from the autoradiogram. Each spot is circumscribed with a sharp pencil or spatula and collected with the device shown in Fig. 11.5, or one that is available from T. W. Winget, Milton, Cambridge, England. With the device shown, the top portion (b) with its sharp ground edge (a) is used to scrape and dislodge the spot. The scraper is connected to a water aspirator pump or vacuum line. A sintered-glass disc of coarse porosity is sealed at (c) to catch the dislodged DEAE-cellulose as it is being sucked up into the tube. When all of the sample spot has been collected, it is washed with 1 or 2 ml of 95% ethanol to remove the urea (with the aid of a slight negative pressure from the vacuum line). The oligonucleotide is then collected by elution with 0.5 ml or less of the TEA-HCO_3 solvent prepared before.

Sequence Analysis of an Isolated Oligonucleotide. For oligonucleotides from a T_1 RNase hydrolysis mixture, the sequence may most often be

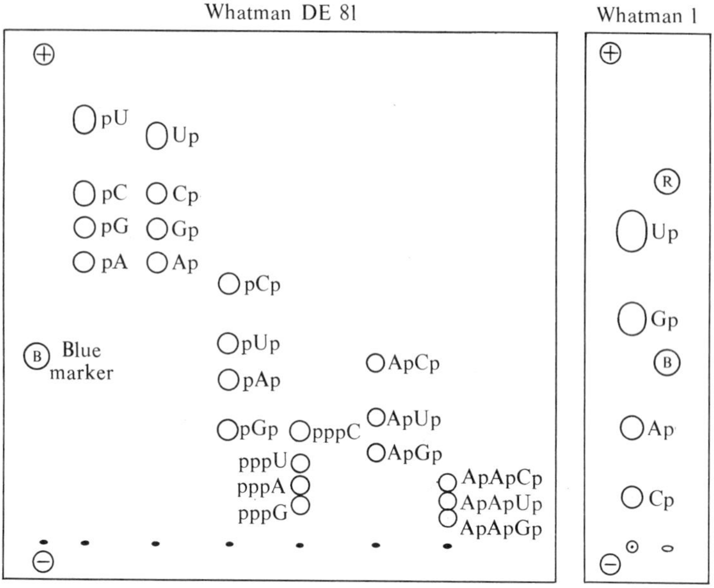

Fig. 11.6. Electrophoretic mobilities of selected nucleotides in 5% acetic acid–0.5% pyridine–0.001 *M* EDTA, pH 3.5.

established after hydrolysis of separate portions with alkali, or pancreatic RNase, or after partial hydrolysis with SPDE. Similarly, for pancreatic RNase oligonucleotides, the sequence may be determined after alkali or T_1 RNase hydrolysis, or partial digestion with SPDE. The procedures are all micro scale and are performed with volumes of 5 to 10 μl, so that they may be spotted immediately on the separation medium. For example, an unknown oligonucleotide from a T_1 RNase digestion is divided into three portions: the first is treated with alkali, the second with pancreatic RNase, and the third is partially digested with SPDE. The products of hydrolysis with alkali are separated by electrophoresis on Whatman 54 paper while the products from the digestion with pancreatic RNase are separated on DEAE-paper. Both electrophoreses are carried out in 0.5% pyridine-5% acetic acid-0.001 *M* EDTA, pH 3.5. Figure 11.6 shows the separations obtained with the products of alkali or pancreatic RNase digestion of Whatman 54 and DEAE-paper, respectively. The determination of base composition in the alkali digestion mixture is accomplished simply by identifying the positions of the radioactive spots with the known positions of the mononucleotides and determining the relative radioactivity of each. The identification of the products from a nuclease digestion may often be accomplished solely by the positions of the smaller oligonucleotides. In cases of doubt, an alkali hydrolysis will establish the base composition which then most often leads to the correct identification. In stubborn cases, a partial digestion with SPDE will reveal the sequence. This is carried out as follows.

SPDE Partial Digestion. The rationale for the use of SPDE partial hydrolysis in sequence analysis is discussed in Chapter 5. The conditions for carrying out this procedure differ slightly, depending on whether the oligonucleotide is obtained from a pancreatic, T_1, or U_2 digestion mixture. This difference is largely due to the different rates of hydrolysis by SPDE of the terminal nucleotides: $U > A = G > C$. Because a C terminal is apt to be encountered in a T_1 or U2 RNase digest, correspondingly longer hydrolysis times with SPDE are required for oligonucleotides from these sources. A typical reaction mixture consists of:

0.3 *M* ammonium succinate, pH 6.5
0.002 *M* EDTA
0.05% Tween 80
0.2 mg SPDE/ml

The dried oligonucleotide is treated with 5 to 10 μl of this mixture for for 30 minutes at 37° in a capillary tube. When the base composition analysis indicates the presence of more than one C residue, the incubation is carried out for 60 minutes.

For oligonucleotides from pancreatic RNase digests, the enzyme concentration may be reduced to 0.1 mg per ml and the incubation maintained for 30 minutes at 37°. At the end of the incubation, the capillary tube is opened and the material in the tube is spotted directly onto a strip of DEAE-paper that is 1 inch wide and 12 inches long. A sample of the undigested oligonucleotide is placed on one side of the digested sample, and on the other side is placed a spot of the combined

Table 11.1. *M* values for partial digestion with phosphodiesterase

	Left End Residue	*M*
A. "T_1" Oligonucleotides + SPDE, ionophoresis at pH 1.0 (Sanger *et al.*, 1965)	C	0.05–0.3
	A	0.5 –1.0
	U	1.5 –2.5
	Left End Residue	*M*
"Pan" Oligonucleotides + SPDE, ionophoresis at pH 1.9	A	0.4–1.1
	G	1.5–2.5
	Right End Residue	*M*
B. "T_1 + PME" Oligonucleotides + VPDE, ionophoresis at pH 3.5 (Brownlee and Sanger, 1967)	C	0.6–1.2
	A	2.1–2.9
	U	1.7–1.9
	G	2.6–4.4

pink and blue markers. The paper strip is mounted on a rack or set into place in the electrophoresis apparatus and then it is wetted with a mixture containing 2.5% formic acid and 8.7% acetic acid. This same solution is used in the reservoirs. Electrophoresis is carried out at about 30 volts per cm until the blue marker moves about one-third of the total distance. The power is turned off and the paper is lifted off carefully and allowed to dry. The positions of the colored markers are marked with radioactive ink and a radioautograph is made of the strip. The distances of all of the spots from the origin are measured and the sequence identification made according to the values shown in Table I. In cases of doubt, the spots may be eluted and a base composition determined after hydrolysis with alkali. This is similar to the procedure suggested by Holley *et al.* (1964) for the determination of sequence through use of VPDE. As a matter of fact, VPDE partial hydrolysis may also be used in the procedure above with the appropriate changes in solution and conditions (Brownlee and Sanger, 1967). For tetramers and some pentamers, it is possible to deduce the complete sequence by application of this kind of analysis of the partial digestion products of SPDE. The longer oligonucleotides may require the combined information from nuclease digests as well.

One source of difficulty in interpreting the results occurs when the oligonucleotide sample consists of a mixture of oligonucleotides. In some cases, where there is a mixture of only two oligonucleotides, the results are interpretable. In others, the oligonucleotides must be separated before analysis with SPDE will yield the correct sequence.

FINGERPRINT PATTERNS FROM NONRADIOACTIVE NUCLEIC ACIDS

Not all nucleic acids can be biosynthetically labeled with a sufficiently high specific activity to permit convenient application of the fingerprint method. As an answer to the problem of visualizing the digestion products from small quantities of nonradioactive nucleic acids, Székely and Sanger (1969) developed a modification for labeling the free 5′-OH groups of oligonucleotides with γ-^{32}P-ATP and polynucleotide kinase (Richardson, 1965). The labeled oligonucleotides that are produced by this procedure may then be fractionated by the two-dimensional fingerprint method in the same way as described before. In practice, nonradioactive oligonucleotides are produced by the combined action of a specific nuclease and PME. Attachment of the radioactive ^{32}P at the left end yields oligonucleotides with essentially the same mobilities as those present in a digest with nuclease alone. As a result, the fingerprint patterns of the oligonucleotides labeled at the left end (with the right end dephosphorylated) are essentially the same as those produced by a nuclease hydrolysis of uniformly labeled nucleic acid.

The requirements for carrying out this procedure are for γ-^{32}P-ATP and nuclease-free polynucleotide kinase. The γ-^{32}P-ATP is prepared by the method of Glynn and Chappell (1964): this is a straightforward method, and no further comment will be made here. The nuclease-free polynucleotide kinase may be prepared according to Takanami (1967) and Sugiura and Takanami (1967). These directions differ from those of Richardson (1965) in that a low nuclease mutant, *E. coli* A-19, is used instead of *E. coli* B, and two additional fractionations by DEAE-cellulose chromatography are substituted for the final phosphocellulose column chromatography. The preparation consists of two main steps: the growth of *E. coli* A-19 infected with T_4r^+ bacteriophage, and the isolation of the polynucleotide kinase from the phage-infected (but not lysed) cells.

Growth of T_4r^+ bacteriophage infected *E. coli* A-19: A 20-liter culture medium is prepared from:

40 g casamino acids (Difco)
60 g KH_2PO_4
120 g Na_2PO_4
10 g NaCl
39 g NH_4Cl
Water to make 20 liters

This solution is autoclaved and sterile additions are made of

20 ml 0.1 *M* $CaCl_2$
20 ml 1 *M* $MgSO_4$
20 ml 1% gelatin solution
200 ml 20% glucose solution

A subculture of 200 ml of *E. coli* A-19 is made in the same medium and used to inoculate the 20-liter culture. Incubation is carried out at 37° with vigorous aeration. The course of the bacterial growth is followed by the increase in turbidity at 650 mμ. When an absorbance of 0.4 to 0.5 (1.5×10^9 cells per ml) is reached, 20 mg of tryptophan are added. This is followed immediately by a suspension of T_4r^+ bacteriophage at a multiplicity of 5.

After 20 to 22 minutes of incubation at 37°, the culture is chilled by the addition of ice and the infected cells are collected by sedimentation at 12,000 × *g* for 30 minutes. The sedimented cells may be stored at −20° for several months.

In order to prepare an extract containing the enzyme, the cells are suspended in 110 ml of buffer containing 0.05 *M* Tris, pH 7.5, and 0.01 *M* glutathione. The cells are then disrupted by sonication for 5 minutes at the maximum settings. During this step, the temperature should not rise above 20°. The cell debris is sedimented at 17,000 × *g* for 30 minutes and the absorbance of the supernatant at 260 mμ is adjusted to 105 by the addition of 0.05 *M* Tris buffer, pH 7.5. This is fraction 1.

The polynucleotide kinase is precipitated from fraction 1 by the addition of 27.6 ml of 5% streptomycin to 138 ml of extract. The streptomycin is added with stirring over a period of 30 minutes. At the end of this time, the suspension is centrifuged at 17,000 × *g* for

10 minutes and the supernatant is discarded. The pellet is suspended in 138 ml of 0.1 *M* potassium phosphate buffer, pH 7.5, containing 0.002 *M* glutathione. This is fraction 2.

Magnesium acetate (0.5 ml of 1 *M*) is added to fraction 2 and the suspension is incubated at 37° for 1.5 hours. At the end of this time, the solution is chilled and the protein precipitate is removed by centrifugation at 17,000 × *g* for 10 minutes. This is fraction 3.

The enzyme in fraction 3 is precipitated by the addition of 10 g of ammonium sulfate per 100 ml of enzyme solution. The ammonium sulfate is added over a 30-minute period with stirring and the stirring is continued for an additional 20 minutes. The precipitate is collected after centrifugation at 17,000 × *g* for 30 minutes, and then it is dissolved in 10 ml of 0.1 *M* potassium phosphate buffer, pH 7.5, containing 0.002 *M* glutathione. This is fraction 4.

Three DEAE-cellulose columns are prepared, one each of the following sizes: 3 × 10 cm, 1 × 10 cm, and 1 × 5 cm. Each column is washed with a starting buffer that consists of 0.01 *M* potassium phosphate, pH 7.5, and 0.01 *M* mercaptoethanol. Fraction 4 is first dialyzed against starting buffer for 5 hours and then it is diluted fivefold with starting buffer. The diluted enzyme solution is applied to the 3 × 10 cm column and washed in with 200 ml of starting buffer. Elution of the enzyme is accomplished with a buffer that consists of 0.05 *M* potassium phosphate, pH 7.5, and 0.01 *M* mercaptoethanol. Fractions are collected and assayed for polynucleotide kinase activity as described below. The tubes containing the polynucleotide kinase activity are combined to make fraction 5.

Fraction 5 is diluted with 4 volumes of 0.01 *M* mercaptoethanol and applied to the 1 × 10 cm DEAE-cellulose column. The column is washed with 50 ml of starting buffer, and then the enzyme is eluted with 0.05 *M* phosphate, pH 7.5, and 0.01 *M* mercaptoethanol. The fractions are assayed, and the polynucleotide kinase activity is combined, diluted with 0.01 *M* mercaptoethanol as before and rechromatographed on the 1 × 5 cm DEAE-cellulose column. The fractions containing the polynucleotide kinase from the last chromatography are combined and may be used in this form or stabilized by the addition of sufficient bovine serum albumin to make a 0.1% solution.

Substrate for the polynucleotide kinase assay may be obtained by the treatment of 5 mg of commercial calf thymus DNA with 10 units of micrococcal nuclease in 0.05 *M* Tris, 0.005 *M* $CaCl_2$, pH 7.5, for 2 to 3 minutes at 37°. The basic medium for the assay consists of: 0.5 *M* Tris, pH 7.5, 0.1 *M* $MgCl_2$, and 0.05 *M* mercaptoethanol. A typical reaction mixture consists of:

0.05 ml basic medium
0.05 ml DNA substrate (0.1 mg)
0.05 ml of γ-^{32}P-ATP (1 μmole/ml)
0.05 to 0.35 ml enzyme

The reaction mixture is incubated at 37° for 30 minutes. It is then chilled in an ice bath, and 2 ml of cold 5% trichloroacetic acid are added. The mixture is allowed to stand for 5 minutes; then it is filtered on a Millipore filter (25 mm, B-6) and washed with cold 0.1 *M* sodium pyrophosphate six times. The filter containing the precipitated DNA is dried and counted in a scintillation counter. One unit of enzyme is the amount needed for the incorporation of 1 mμmole of ^{32}P in 30 minutes. An alternative assay is given by Székely and Sanger (1969) using dimers under essentially the conditions used for the phosphorylation of nucleic acid digests.

CHAIN LENGTH SEPARATION OF OLIGONUCLEOTIDES

One of the more important developments in nucleic acid sequence analysis was the use by Tomlinson and Tener (1963) of 7 *M* urea in conjunction with DEAE-cellulose chromatography for the separation of oligonucleotides according to chain length. In this system, it has been found that the concentration of salt needed to elute an oligonucleotide is proportional to the logarithm of the charge of the oligonucleotide. With oligonucleotides from homopolymers, separations may be achieved for chain lengths of 20 and beyond. With oligonucleotides from nuclease digests of RNA, the separations are not so clear beyond a chain length of 5 or 6; the resolution is blurred as the base compositional complexity increases. (See Chapter 2 for discussion.)

With radioactive RNA, separations may be conveniently carried out on small columns. As much as 10 mg of RNA digest may be separated on columns of the size 0.5 × 30 cm. The lower limit depends on the method of testing for the emergence of oligonucleotides. Probably the only critical requirement in column chromatography at this dimension is the rate of elution. Probably the rate should not exceed 6 ml per hour, and the total gradient should not be less than 50 column volumes.

PREPARATION OF DEAE-CELLULOSE OR DEAE-SEPHADEX COLUMNS

1. DEAE-cellulose (or DEAE-Sephadex) is suspended in 7 *M* urea containing 2 *M* NaCl. The mixture is titrated with HCl until the pH remains constant at 3. After 30 minutes, the slurry is washed with 3 volumes of unbuffered 7 *M* urea-2 *M* NaCl solution. This washing may be carried out in a large column or with the aid of an aspirator-filter. The DEAE-cellulose is then washed in 7 *M* urea, the excess liquid is removed on an aspirator-filter, and the filter cake is suspended in starting buffer.
2. The DEAE-cellulose is poured as a slurry into a column that has been filled halfway with starting buffer. The addition of the slurry is continued until the desired column bed height is reached. The column is then washed with at least 3 column volumes of starting buffer.
3. The sample of oligonucleotides, dissolved usually in 7 *M* urea or starting buffer is loaded on the column, washed in twice with small amounts of 7 *M* urea or starting buffer, and the elution is begun.

A generally useful starting buffer for chromatography with DEAE-cellulose is 7 *M* urea-0.01 *M* Tris, pH 7.5 to 8. For DEAE-Sephadex, a suitable starting buffer is 7 *M* urea-0.15 *M* NaCl-0.01 *M* Tris, pH 7.5 to 8. The difference in the starting buffers as well as the slope of the gradient is related to the fact that the ion-exchange capacity of DEAE cellulose is usually ⅓ of the capacity of DEAE-Sephadex.

PREPARATION OF MACALOID GEL

Macaloid powder is obtained from the Baroid Division, National Lead Co., Houston, Texas. Macaloid powder (30 g) is added with mixing to 500 ml of water at 90 to 100°: the addition is made in a high-shear mixer such as a Waring Blendor. The blending is continued until a homogeneous thixotropic gel is formed. Concentrated HCl is added dropwise until the pH drops to 2 as estimated by pH paper. The $CaCO_3$ present in the crude Macaloid will dissolve slowly, raising the pH. HCl is added at intervals until the pH remains constant at 2. The gel is centrifuged at 25,000 *g* for 15 minutes and the supernatant fluid is discarded. The residue is mixed with 50 ml of 0.2 *M* Tris base and then dispersed in hot water to a final volume of 500 ml. The dispersal of the gel is performed in a high-shear mixer. The gel is next titrated with continuous stirring to a pH between 7 and 7.5 with 1 *M* KOH. The mixture is centrifuged as before and the supernatant is discarded. The residue is suspended in a minimum volume (300 to 400 ml) of 0.02 *M* Tris, pH 7.2. The gel is then placed in small flasks in portions of 25 g each and autoclaved for 20 minutes. The capped flasks are stored at 4° until ready for use.

PREPARATION OF BENTONITE

Preparation technique is that of Fraenkel-Conrat *et al.*, 1961. Bentonite is obtained from the International Minerals and Chemical Corp. of Skokie, Ill.

1. Bentonite (100 g) is suspended in 1500 ml of water, with stirring, until a smooth suspension is obtained. The suspension is centrifuged at 2000 *g* for 15 minutes. The supernatant is retained and the residue is discarded.
2. The supernatant is then centrifuged in a Spinco No. 21 rotor at 9000 rpm for 20 minutes. The supernatant from this is discarded and the residue is suspended in 500 ml of 0.1 *M* EDTA, pH 7 to 7.5, and stirred for 48 hours.
3. The smooth suspension is centrifuged at 2000 *g* for 15 minutes and the residue is discarded. The supernatant is centrifuged at 12,000 *g* for 20 minutes and the supernatant is discarded.
4. The residue is suspended in 300 ml of 0.01 *M* acetate buffer, pH 6, and stirred until a smooth mixture suspension is obtained. The suspension is centrifuged at 12,000 *g* for 20 minutes; then the supernatant is discarded and the residue is suspended in 150 ml of 0.01 *M* acetate buffer, pH 6. The bentonite concentration is determined by measuring the weight of dry solid in a known volume.

GEL ELECTROPHORESIS OF OLIGONUCLEOTIDES

The fractionation of oligonucleotides with molecular weights less than 10^4 has been achieved by electrophoresis on 15% polyacrylamide gels (Gould *et al.*, 1969). The oligonucleotides are separated mainly on the basis of molecular size, the smallest molecules migrating farthest through the gel. In this way it is somewhat similar to the chain length separations by column chromatography except that, here, the resolution of the longer chain lengths is relatively better than with column chromatography. This is illustrated in Fig. 11.7, which is a comparison of the positions of oligonucleotides from a T_1 RNase hydrolysis of TMV RNA in both gel electrophoresis and column chromatography. It may be seen that, on the gel, oligonucleotides of chain length 8 and above are resolved but not those with chain lengths below. With the column, oligonucleotides with chain lengths below 6 are resolved and then become less so for chain lengths above this number. It seems, therefore, that gel electrophoresis may also serve as a valuable means for fractionation of oligonucleotides above the chain lengths of 8.

The following procedure utilizes the discontinuous buffer system in order to take advantage of the zone sharpening effect.

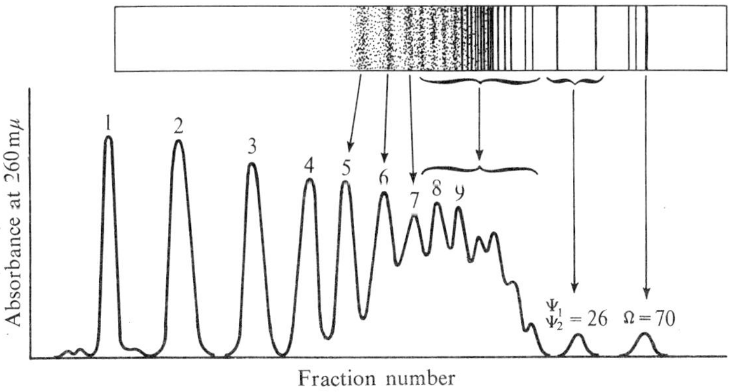

Fig. 11.7. Correlation between chain-length chromatography and disc electrophoresis.

Materials. The materials required are the same as described before for the polyacrylamide gel electrophoresis of tRNA extracts. Buffers A, B, and C are made up as specified in Chapter 10. The gel solutions needed are made up as follows for the 15% running gel:

Solution 1 60 g acrylamide
0.4 g bisacrylamide
14 mg $K_3Fr(CN)_6$
Water to make 100 ml

Solution 2 50 ml buffer A
0.23 ml TEMED
Water to make 50 ml

Solution 3 0.14 g ammonium persulfate
water to make 100 ml
Solution 3 is degassed with the aid of a water aspirator pump.

Method Mix 2 parts of solution 1 with 1 part of solution 2 and 1 part of water. Degas. This is solution 4.
Mix equal parts of solution 3 with solution 4 and use to load plastic gel tubes.

The solution prepared above is used to load Lucite or polyethelene tubes as described before. The plastic tubes are used to permit easier removal of the gel after electrophoresis.

The 10% spacer gel is made according to the directions given for the 10% running gel except that, for the spacer gel, buffer B is used in solution 2 instead of buffer A. The remainder of the procedure is carried out as specified before. The only disadvantage is that oligonucleotides smaller than 5-mers are washed out of the gel during the staining and destaining operations. Otherwise, the positions of the bands may be visualized and the radioactive oligonucleotides may be recovered from sections of the gel as described before.

DESALTING PROCEDURES

The problem of removing salts and urea from oligonucleotide fractions is often encountered in sequence analysis. For oligonucleotides up to a chain length of 4, absorption methods may be used. One

Table 11.2. Dialysis of $(Ap)_n$ oligonucleotides[a]

Oligonucleotides	Percent Remaining	Oligonucleotides	Percent Remaining
Ap	1.2	$(Ap)_4$	68
$(Ap)_2$	9	$(Ap)_5$	96
$(Ap)_3$	32	$(Ap)_6$	94

[a] $(Ap)_n$ oligonucleotides isolated from column chromatography (DEAE-cellulose-7M urea) of alkaline hydrolysis of poly A. Each fraction dialyzed against four changes of water at 4° for 24 hours.

convenient method is the absorption of the oligonucleotide on activated charcoal. This is followed by several washes with water and then elution with a solution containing 60% ethanol and 0.1 *M* ammonium hydroxide (Mandeles and Kammen, 1966.) The ethanol-ammonia is removed by evaporation *in vacuo* and the moist residue is freeze-dried. As many as 1000 A_{260}-cm^2 units of oligonucleotides may be processed with two or three filtration setups in a few hours. The yields vary from 90% for monomers to about 70% for trimers. This procedure is not recommended for tetramers because of poor yields.

Another method depends on the reabsorption of the oligonucleotides from a column chromatographic fractionation onto fresh DEAE-cellulose or DEAE-Sephadex. This is accomplished by diluting the oligonucleotide solution 1:5 with water and loading the diluted solution on a column usually no more than one-half the size of the column originally used in the chain length fractionation. After the sample is loaded, the column is washed extensively with water to remove urea and salts. The oligonucleotides may then be eluted with a solution of 2 *M* TEA-HCO_3. More efficient elution of oligonucleotides may be achieved with a linear gradient from 0 to 2 *M* TEA-HCO_3. The concentration of TEA-HCO_3 required for tetramers is less than 1 *M*; the shorter oligonucleotides may be eluted at lower concentrations. The TEA-HCO_3 is removed by evaporation (freeze-drying). In some instances, NH_4HCO_3 has been used for the same purpose, but TEA-HCO_3 has been found to be more easily removed.

For pentamers and longer oligonucleotides, dialysis has been very effective. As shown in Table 11.2, as much as 90% of $(Ap)_5$ is retained after dialysis against water with several changes. At the same time, almost all of the mononucleotide Ap is lost. Precautions against degradation during dialysis are similar to those taken during the other manipulations with nucleic acids. Dialysis tubing is boiled briefly in dilute EDTA-SDS solutions and rinsed with a dilute bentonite suspension. Bentonite may also be added to the dialysis solution and later removed by centrifugation at 17,000 *g* for 30 minutes. Small-volume dialysis may be accomplished with commercial dialyzers offered by the Chemical Rubber Co. Large volumes (from 200 ml to liter quantities) may be dialyzed and concentrated rapidly in a counterflow device manufactured by Omnivector Instruments, Freehold, New Jersey.

Desalting as well as concentration of an oligonucleotide solution may be accomplished also with the aid of the Amicon Diaflo apparatus. Membranes are available that permit the passage under pressure of solutions containing salt and urea but not larger molecules such as trimers. In practice, the oligonucleotide solution is reduced in volume to about a 0.1 volume, diluted with water, and filtered under pressure to reduce the volume again. This is repeated until an acceptably low salt and urea concentration is reached in a suitably small volume of solution.

FORMATION OF CYTIDYLIC ACID-TERMINATED OLIGONUCLEOTIDES

Oligonucleotides with cytidylic acid at the right end may be produced from nucleic acids by either of two methods:

1. Use of a C-specific nuclease that may be isolated from partially purified bacterial alkaline phosphatase (phosphomonoesterase) preparations (Anderson and Carter, 1965).

2. Action of pancreatic RNase on nucleic acid containing chemically modified uracil residues (Lee *et al.*, 1965; Brownlee *et al.*, 1968) followed by the removal of the modifying substituents.

Experience with the C-specific nuclease from crude alkaline phosphatase preparations indicates that it is suitable for use with oligonucleotides and the smaller nucleic acids. Chain length fractionation of the C-nuclease products from nucleic acids as large as TMV RNA does not produce patterns equivalent in resolution to those obtained with either pancreatic or T_1 RNase. The reason for this is not clear. One possibility is that not all of the theoretically susceptible bonds have been hydrolyzed. Another is that some of the C-terminated oligonucleotides remain in the intermediate 2′,3′-cyclic phosphate form rather than the 3′-phosphomonoester. Either of these possibilities would tend to increase the complexity of the pattern and thus decrease the resolution. With smaller oligonucleotides, the number of possible products is limited and thus the results are more easily interpreted.

The use of pancreatic RNase to produce oligonucleotides terminating in a cytidylic acid residue depends on the prior reaction of the nucleic acid with 1-cyclohexyl-3-(2-morpholinyl-(4)-ethyl)-carbodiimide metho-*p*-toluenesulfonate (CMCT). At pH 9, CMCT reacts mainly with uracil and guanine residues at positions 2 and 4, respectively. A schematic of the reaction with uracil is shown in Fig. 11.8. The substituted nucleic acid is then hydrolyzed with pancreatic RNase. Because

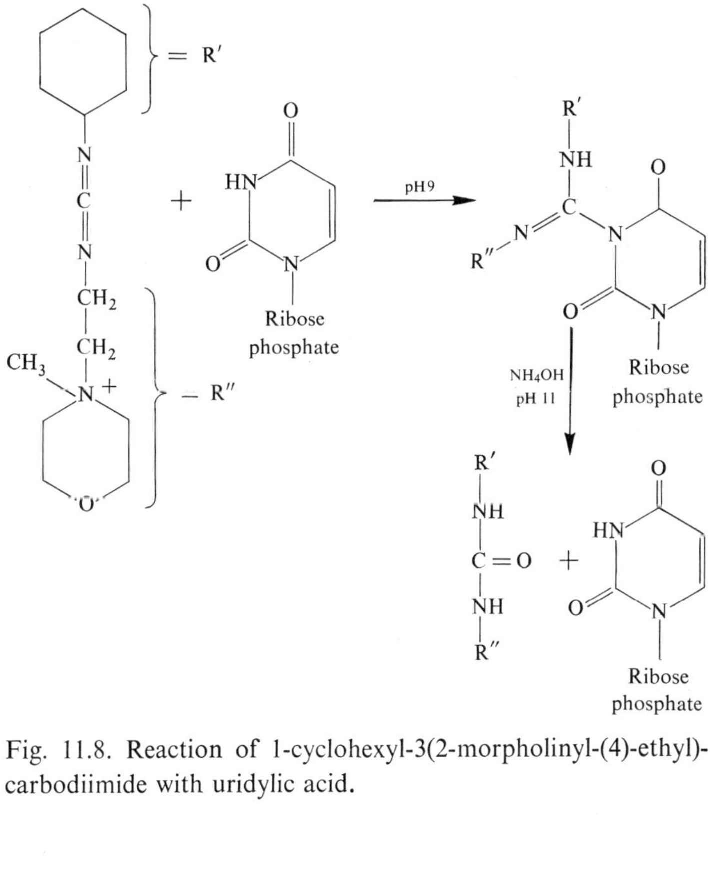

Fig. 11.8. Reaction of 1-cyclohexyl-3(2-morpholinyl-(4)-ethyl)-carbodiimide with uridylic acid.

the substituted uridylic acid residues are resistant to hydrolysis, only the bonds adjacent to the cytidylic acid residues are cleaved. After hydrolysis is complete, the nuclease is removed with bentonite or by reduction with mercaptoethanol and digestion with the proteolytic enzyme pronase. The CMCT derivative is then removed from the C-terminated oligonucleotides by incubation with ammonium hydroxide at pH 11 for a short time.

When this procedure is used with oligonucleotides or small nucleic acids (Brownlee *et al.*, 1968), not all of the products contain cytidylic acid exclusively at the right terminus. However, the number of products is limited and the results are easily interpreted. With TMV RNA, the oligonucleotides produced by this procedure show the same complexity as those produced by the C-specific nuclease.

PREPARATION OF C-SPECIFIC RNase

A suitable starting material may be obtained as bacterial alkaline phosphatase BAP-SF from Worthington Biochemicals. This may also be obtained by the directions of Garen and Levinthal (1960). A summary of these directions follows.

1. *Escherichia coli* B (20 liters) is grown on phosphate-limiting medium.

2. The cells are harvested by centrifugation, suspended in 100 ml of 0.1 *M* Tris, pH 7.4, and incubated at 82° for 15 minutes.

3. The suspension is first dialyzed against 0.1 *M* Tris, pH 7.4, for 20 hours at 4°, then centrifuged at 20,000 × *g* for 20 minutes, and the supernatant is saved. The pellets are washed three times with 10-ml portions of 0.1 *M* Tris, pH 7.4, and the supernatants are combined. The supernatants are centrifuged at 20,000 × *g* for 1 hour, and the pellet is discarded.

4. The supernatant is then fractionally precipitated with saturated ammonium sulfate. The fraction between 0.45 and 0.65 saturation is dialyzed against 0.1 *M* Tris, pH 7.4, and then incubated at 82° for 15 minutes as before. The enzyme solution is clarified by centrifugation at 20,000 × *g* for 1 hour and the enzyme is precipitated by adjusting the ammonium sulfate concentration to 0.85 saturation.

5. The precipitate is centrifuged at 17,000 × *g* for 30 minutes and the pellet is suspended in a minimum volume of 0.65 saturated ammonium sulfate. At this stage, the preparation is equivalent to the Worthington Biochemicals alkaline phosphatase BAP-SF.

The alkaline phosphatase (either Worthington BAP-SF or the preparation described above) is diluted with water to a final concentration of 2 mg of of protein per milliliter. This solution is chilled to 4° and mixed with 0.2 volume of chilled 1 *M* $HClO_4$. The precipitate is removed by centrifugation at 17,000 × *g* for 15 minutes and the supernatant is neutralized with chilled 1 *M* KOH. The mixture is allowed to stand in the cold for 2 hours and then the $KClO_4$ precipitate is removed by centrifugation at 17,000 × *g* for 10 minutes. The supernatant is dialyzed overnight against 4 liters of water and then freeze-dried.

Enzyme activity is assayed by measuring the formation of acid-soluble products with RNA as a substrate. The specificity may be checked through use of poly-C, poly-A, poly-U as substrates in control reactions. A typical assay mixture consists of

(a) 0.2 mg of RNA in 2.9 ml of 0.04 *M* Tris, pH 8.0;
(b) C-specific RNase in 0.1 ml of water. The amount of enzyme used should yield an increase in absorbance of no more than 1 unit.

After incubation at 37° for 10 minutes, the reaction is terminated by the addition of 1 ml of 1 *M* $HClO_4$. The mixture is chilled, centrifuged, and the absorbancy $A_{260\ m\mu}$ is measured. This measurement is made against a blank prepared by adding 1 ml of 1 *M* $HClO_4$ to the RNA solution prior to the addition of the enzyme. One enzyme unit produces an increase in the $A_{260\ m\mu}$ of 1.0 for a 10-minute reaction time. This unit of activity may be compared with similar units for pancreatic and T_1 RNase in order to define reaction conditions for the complete or partial hydrolysis of nucleic acids.

CMCT AND THE FORMATION OF C-TERMINAL OLIGONUCLEOTIDES

The following procedure is essentially similar to the one described by Brownlee *et al.* (1968), except that bentonite is used to remove pancreatic RNase instead of destroying the nuclease by digestion with pronase.

Materials

Buffer A 0.02 *M* Tris-0.002 *M* EDTA, pH 8.95
Buffer B 0.02 *M* Tris-0.002 *M* EDTA, pH 7.5
Buffer C 1.0 *M* Na acetate, pH 5.0
Pancreatic RNase (Worthington RAF)
Bentonite prepared as described in this chapter
CMCT 1-cyclohexyl-3-(2-morpholinyl-(4)-ethyl)-carbodiimide metho-*p*-toluenesulfonate obtained from Aldrich Chemical Company, Milwaukee, Wisconsin.

Method. In a small, siliconized Corex centrifuge tube, mix 20 μl of buffer A containing 1 to 100 μg of ^{32}P-RNA with 20 μl of buffer A containing 2 mg of CMCT. Incubate at 37° for 16 to 18 hours. At the end of the incubation, add 5 μl of buffer C and 200 μl of 95% ethanol, and allow to stand at −20° for 4 hours. The suspension is then centrifuged at 17,000 × *g* for 30 minutes and the supernatant is drawn off with a micropipet.

The residue is dissolved in 50 μl of buffer B and to this is added 50 μl of pancreatic RNase solution in buffer B (one unit of enzyme activity for every 20 μg of ^{32}P-RNA). The reaction mixture is incubated for 4 hours at 40°. At the end of the incubation, bentonite is added to adsorb the nuclease. Approximately 2 μg of bentonite are used for each unit of enzyme activity. The mixture is chilled for several hours and then centrifuged at 17,000 × *g* for 30 minutes to sediment the bentonite-nuclease complex. The supernatant is removed with a micropipet and then mixed with 2 volumes of concentrated ammonium hydroxide. This mixture is allowed to stand at room temperature for 2 hours and then freeze-dried. The residue is taken up in 5 to 20 μl of water for two-dimensional electrophoresis.

REFERENCES

Adams, J. M., Jeppesen, P. G. N., Sanger, F., and Barrell, B. G. (1969). *Nature*, **223,** 1009.
Anderson, J. H., and Carter, C. E. (1965). *Biochemistry*, **4,** 1102.
Arima, T., Uchida, T., and Egami, F. (1968a). *Biochem. J.*, **106,** 601.
Arima, T., Uchida, T., and Egami, F. (1968b). *Biochem. J.*, **106,** 609.
Brownlee, G. G., Sanger, F., and Barrell, B. G. (1968). *J. Mol. Biol.*, **34,** 379.

Brownlee, G. G., and Sanger, F. (1969). *Eur. J. Biochem.*, **11**, 395.
Fellner, P., and Ebel, J. P. (1970). *Nature*, **225**, 1131.
Fellner, P., Ehresmann, C., and Ebel, J. P. (1970). *Nature*, **225**, 26.
Fraenkel-Conrat, H., Singer, B., and Tsugita, A. (1961). *Virology*, **14**, 54.
Garen, A., and Levinthal, C. (1960). *Biochim. Biophys. Acta*, **38**, 470.
Glynn, I. M., and Chappell, J. B. (1964). *Biochem. J.*, **90**, 147.
Gould, H. J., Pinder, J. C., Matthews, H. R., and Gordon, A. H. (1969). *Anal. Biochem.*, **29**, 1.
Gupta, S. L., Chen, J., Schaefer, L., Lengyel, P., and Weissman, S. M. (1970). *Biochem. Biophys. Res. Commun.*, **39**, 883.
Holley, R. W., Madison, J. T., and Zamir, A. (1964). *Biochem. Biophys. Res. Commun.*, **17**, 389.
Lee, J. C., Ho, N. W. Y., and Gilham, P. T. (1965). *Biochim. Biophys. Acta*, **95**, 503.
Lloyd, D. A., and Mandeles, S. (1970). *Biochemistry*, **9**, 932.
Mandeles, S., and Kammen, H. O. (1966). *Anal. Biochem.*, **17**, 540.
Richardson, C. C. (1965). *Proc. Natl. Acad. Sci.*, **54**, 158.
Rushizky, G. W., Bartos, E. M., and Sober, H. A. (1964). *Biochemistry*, **3**, 626.
Rushizky, G. W., Greco, A. E., Hartley, R. W., and Sober, H. A. (1963). *Biochem. Biophys. Res. Commun.*, **10**, 311.
Rushizky, G. W., and Knight, C. A. (1960a). *Virology*, **11**, 236.
Rushizky, G. W., and Knight, C. A. (1960b). *Proc. Natl. Acad. Sci.*, **46**, 945.
Rushizky, G. W., Knight, C. A., and Sober, H. A. (1961). *J. Biol. Chem.*, **236**, 2732.
Rushizky, G. W., and Sober, H. A. (1962). *J. Biol. Chem.*, **237**, 834.
Sanger, F., Brownlee, G. G., and Barrell, B. G. (1965). *J. Mol. Biol.*, **13**, 373.
Sanger, F., and Tuppy, H. (1951). *Biochem. J.*, **49**, 463, 481.
Sato, K., and Egami, F. (1957). *J Biochem.*, **44**, 753.
Sugiura, M., and Takanami, M. (1967). *Proc. Natl. Acad. Sci.*, **58**, 1595.
Székely, M., and Sanger, F. (1969). *J. Mol. Biol.*, **43**, 607.
Takahashi, K. (1961). *J. Biochem.*, **49**, 1.
Takahashi, K. (1962). *J. Biochem.*, **51**, 95.
Takanami, M. (1967). *J. Mol. Biol.*, **23**, 135.
Tomlinson, R. V., and Tener, G. M. (1963). *Biochemistry*, **2**, 697.
Volkin, E., and Cohn, W. E. (1953). *J. Biol. Chem.*, **197**, 521.

12.

SUMMARY OF NUCLEIC ACID SEQUENCES AND STRUCTURE

As more and more sequences are reported, the evidence seems to indicate that messenger RNA contains more information than that required for the strict colinear specification of amino acid sequences in proteins. It appears that, in mRNA, some optimal secondary structure exists whose function is not yet clear. The secondary structure (through base complementarity and hydrogen bonding) of tRNA may have a function in terms of its participation in specific enzyme reactions. The secondary structure of rRNA and even some virus RNA may fulfill the requirements for efficient packing into a small organelle. But what is the reason for extensive (greater than would be expected for random codon sequences) secondary structure in a nucleic acid whose primary purpose is to be translated in a linear fashion? The answers to this question and others are sure to come as further details about the structure nucleic acids and the mechanism of protein synthesis are obtained.

The genetic dictionary is given in Fig. 12.1, and a summary of the structures of the known nucleic acid sequences (to date) is shown in Figs. 12.2 through 12.22. These structures are hypothetical, and they illustrate the extensive possibilities for base complimentary secondary structure. The regularity seen in the tRNAs has argued for its acceptance as tRNA structure and indirectly for the validity of secondary structure in other nucleic acids.

Amino Acid	Codon	Amino Acid	Codon
H– glycine	GGN	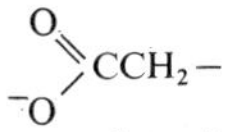 glutamic acid	GAY
CH_3– alanine	GCN	 aspartic acid	GAR
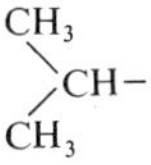 valine	GUN	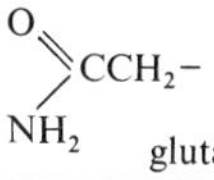 glutamine	AAY
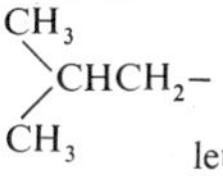 leucine	CUN UUR	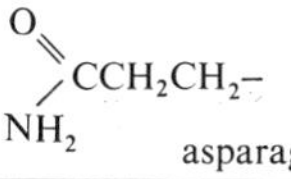 asparagine	CAR
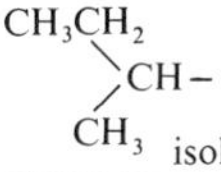 isoleucine	AUY AUA	$H_2NCH_2CH_2CH_2CH_2$– lysine	AAR
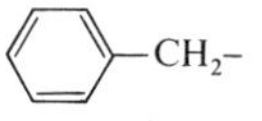 phenylalanine	UUY	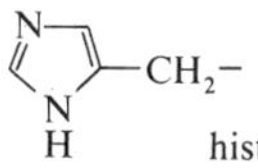 histidine	CAY
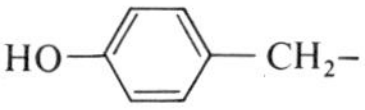 tyrosine	UAY	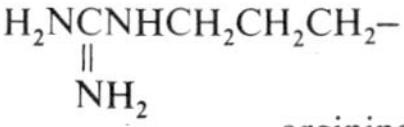 arginine	CGN AGR
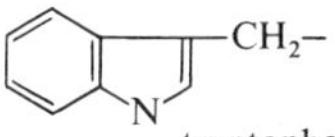 tryptophan	UGG	$HOCH_2$– serine	AGY UCN
$HSCH_2$– cysteine	UGY	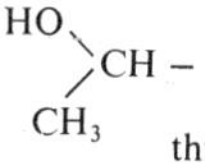 threonine	ACN
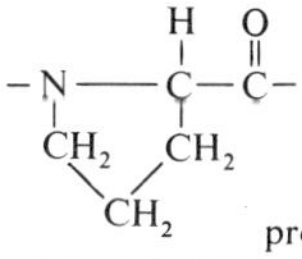 proline	CCN	$CH_3SCH_2CH_2$– methionine	AUG
Start ≡ formyl methionine	AUG GUG	Stop	UAR UGA

Fig. 12.1. The genetic dictionary.

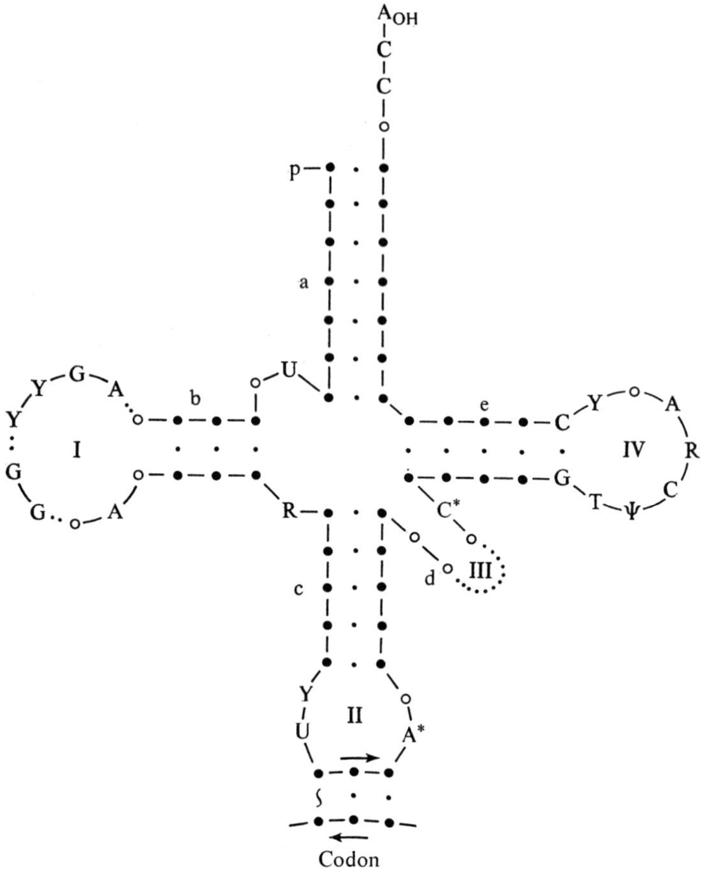

Fig. 12.2. Generalized structure of tRNA.

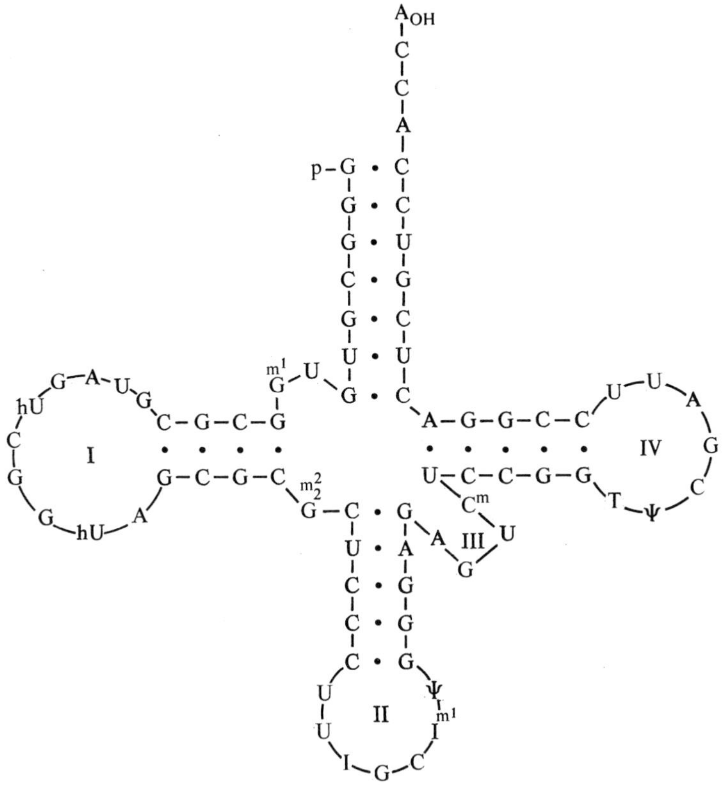

Fig. 12.3. tRNA^{ala} from yeast (Holley *et al.*, 1965).

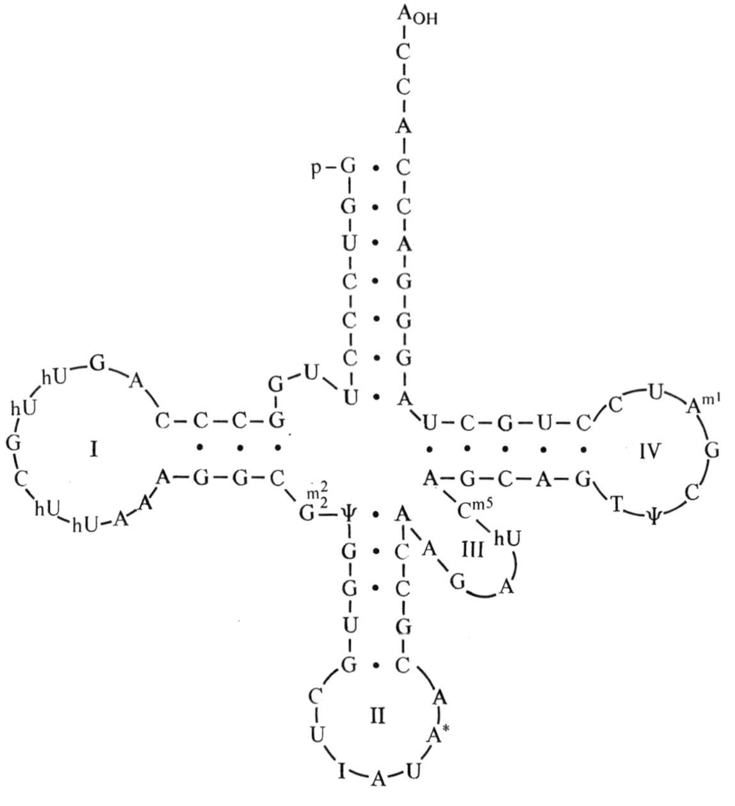

Fig. 12.4. tRNA^{ile} from *T. utilis* (Takemura *et al.*, 1969).

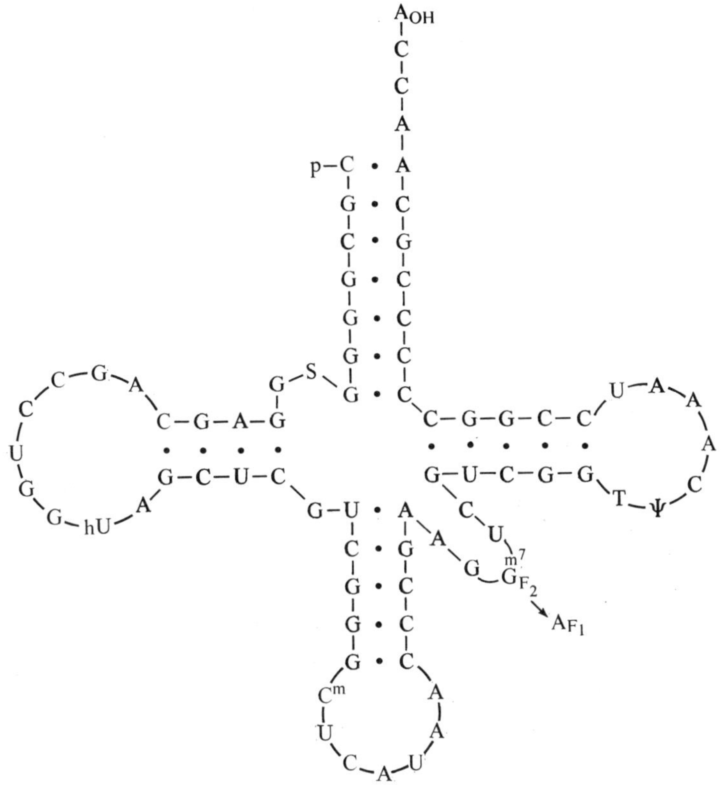

Fig. 12.5. tRNA$^{\mathrm{met}}_{\mathrm{F_2\ and\ F_1}}$ from *E. coli* (Dube *et al.*, 1968).

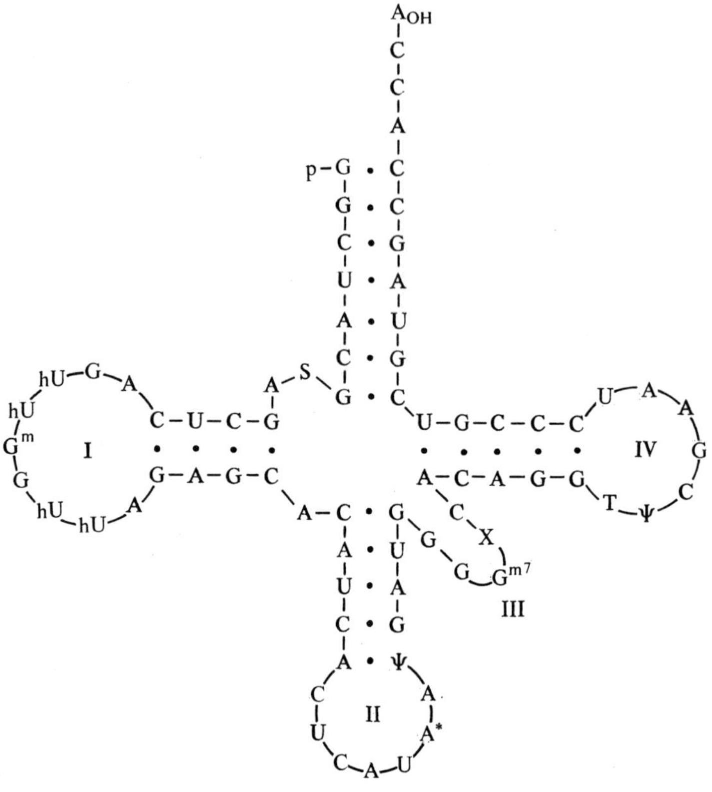

Fig. 12.6. tRNA$^{met}_{M}$ from *E. coli* (Cory *et al.*, 1968).

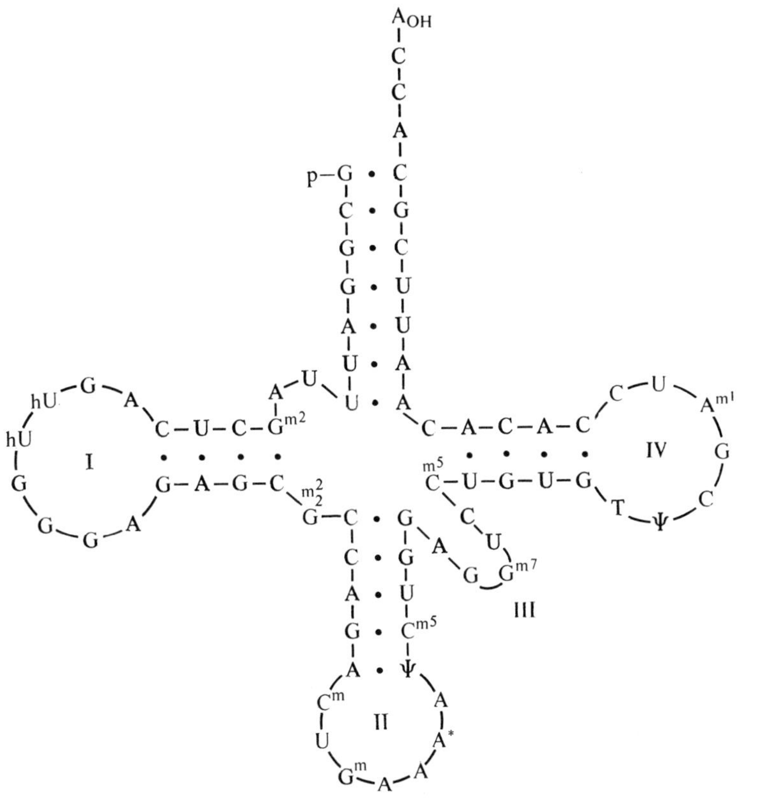

Fig. 12.7. tRNA^{phe} from bakers' yeast (RajBhandary and Chang, 1968).

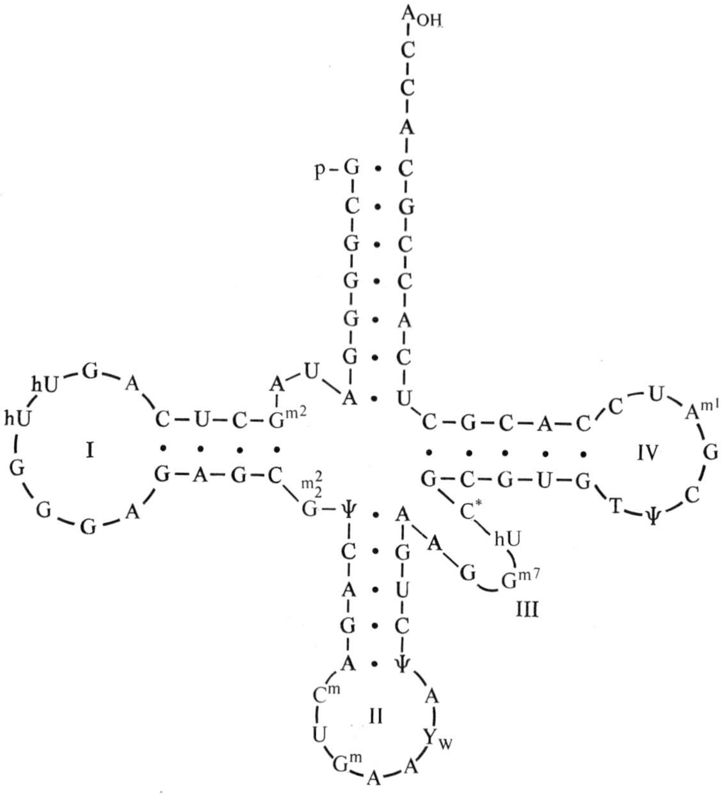

Fig. 12.8. tRNA^{phe} from wheat germ (Dudock *et al.*, 1969).

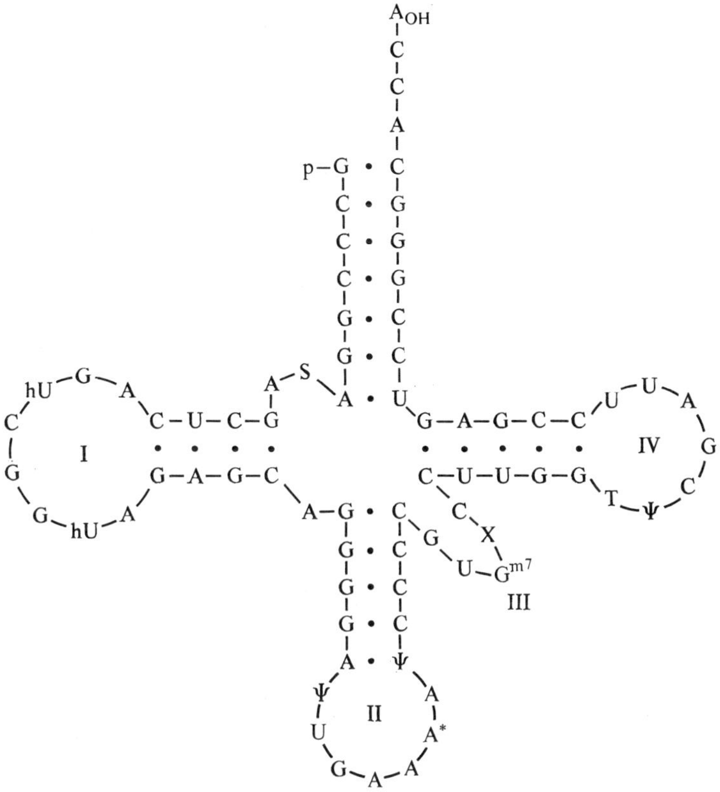

Fig. 12.9. tRNA^{phe} from *E. coli* (Barrell and Sanger, 1969).

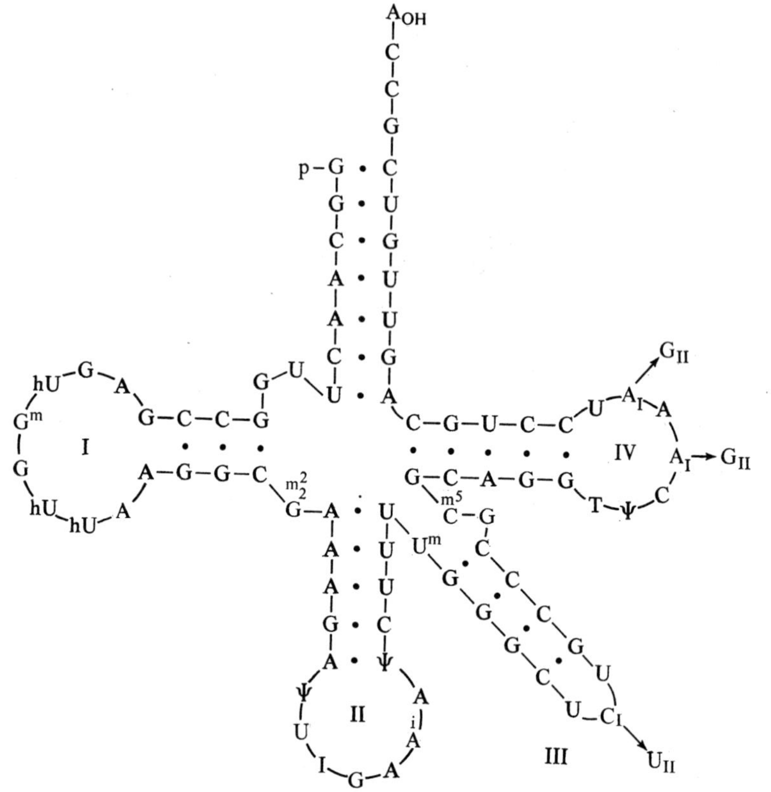

Fig. 12.10. tRNA$^{ser}_{I\ and\ II}$ from brewers' yeast (Zachau *et al.*, 1966).

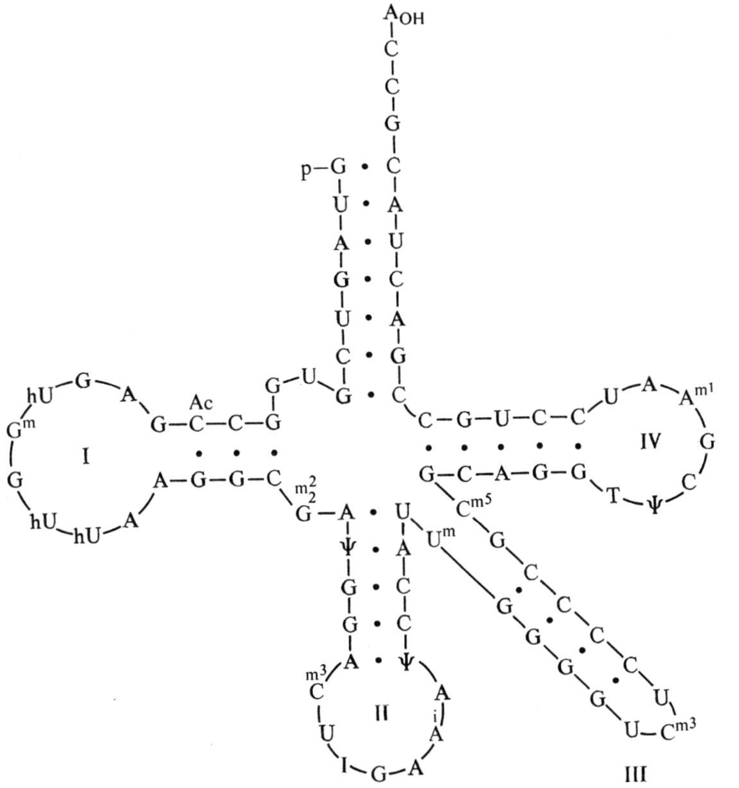

Fig. 12.11. tRNA^{ser} from rat liver (Staehelin *et al.*, 1968).

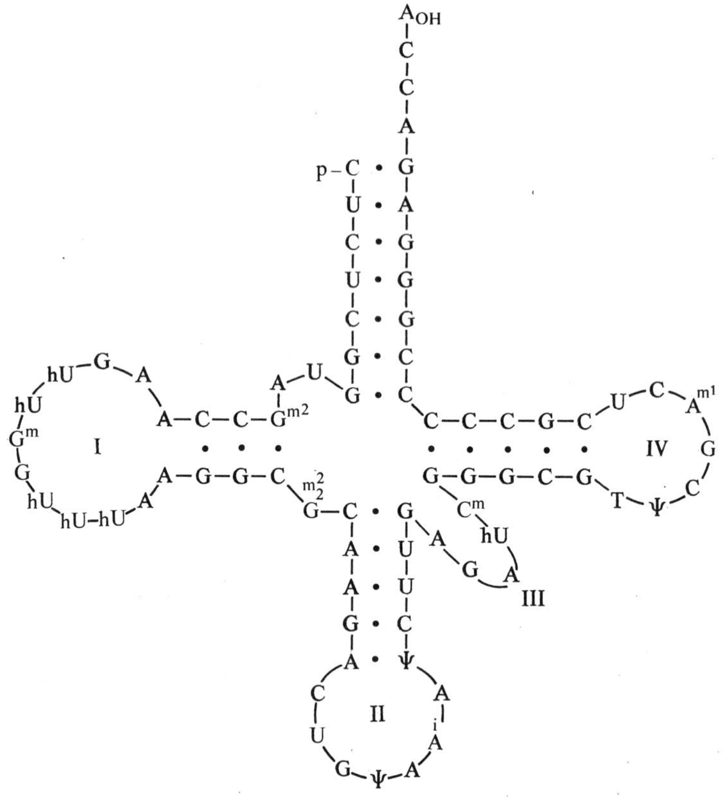

Fig. 12.12. tRNA^{tyr} from bakers' yeast (Madison *et al.*, 1966).

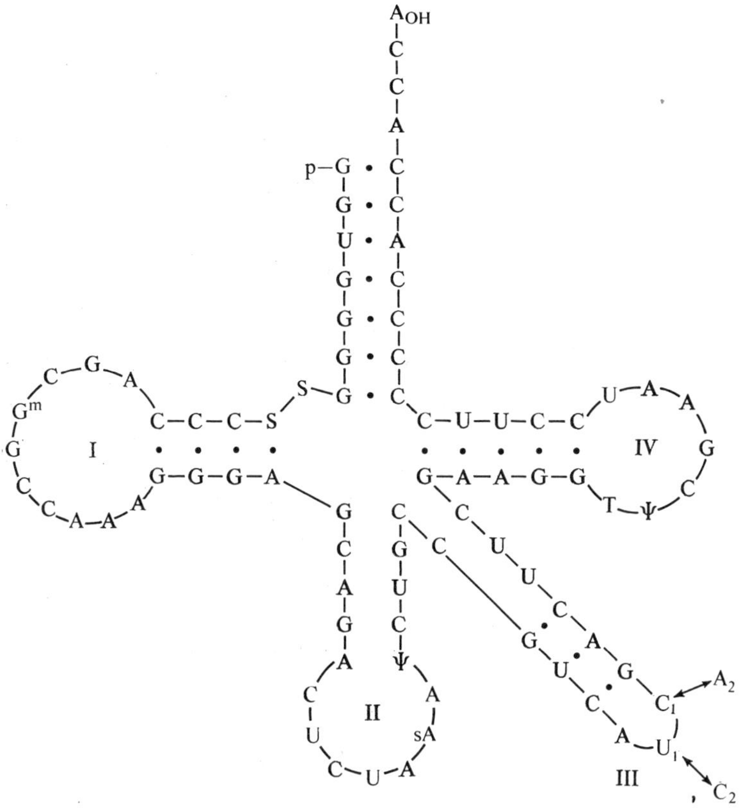

Fig. 12.13. $tRNA^{tyr}_{Su^{+}III}$ from *E. coli*, ϕ80 (Goodman *et al.*, 1968).

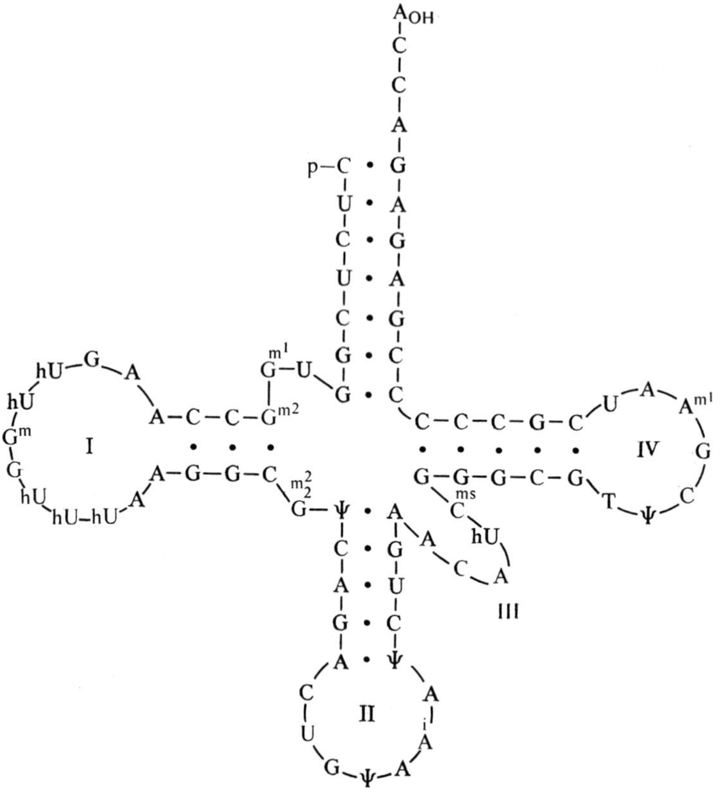

Fig. 12.14. tRNA^{tyr} from *T. utilis* (Hashimoto *et al.*, 1969).

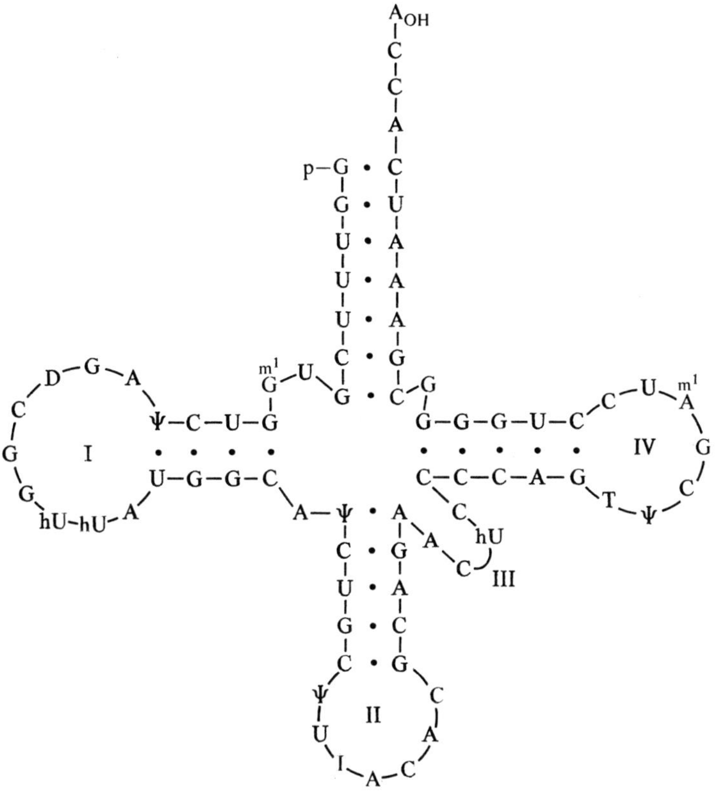

Fig. 12.15. tRNA^{val} from bakers' yeast (Baev *et al.*, 1967).

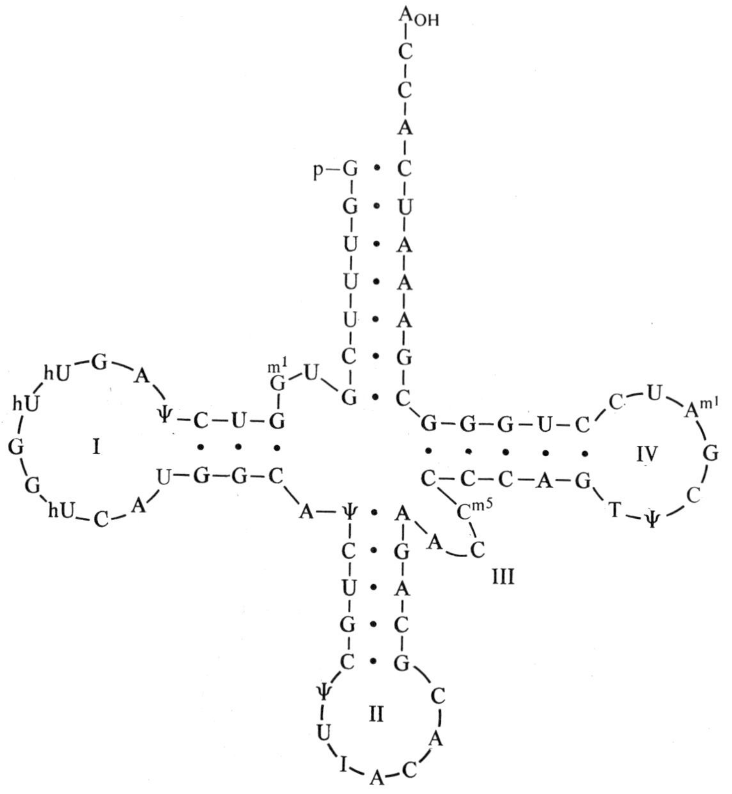

Fig. 12.16. tRNA$_1^{val}$ from *T. utilis* (Mizutani *et al.*, 1968).

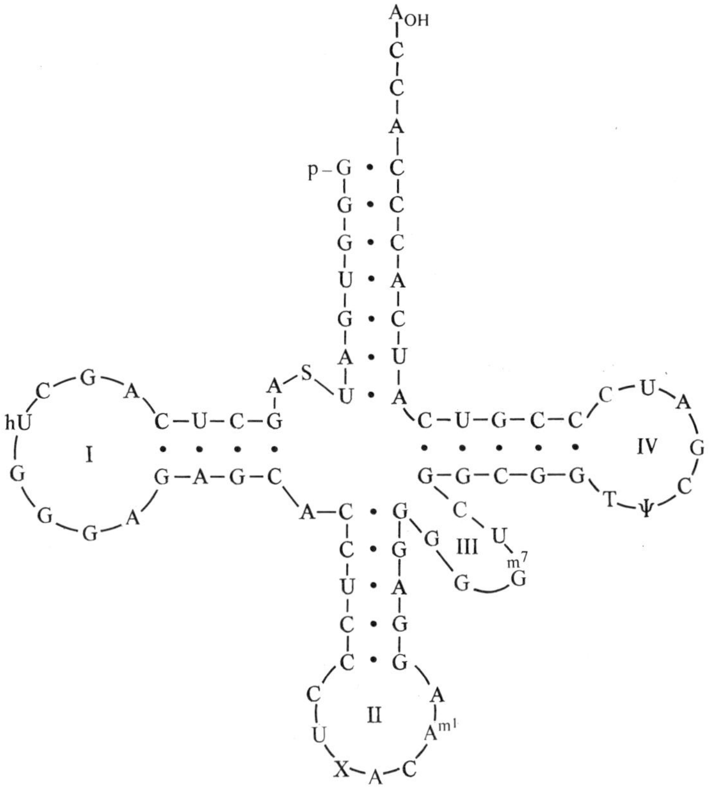

Fig. 12.17. tRNA$_1^{val}$ from *E. coli* B (Yaniv and Barrell, 1969; Harada *et al.*, 1969).

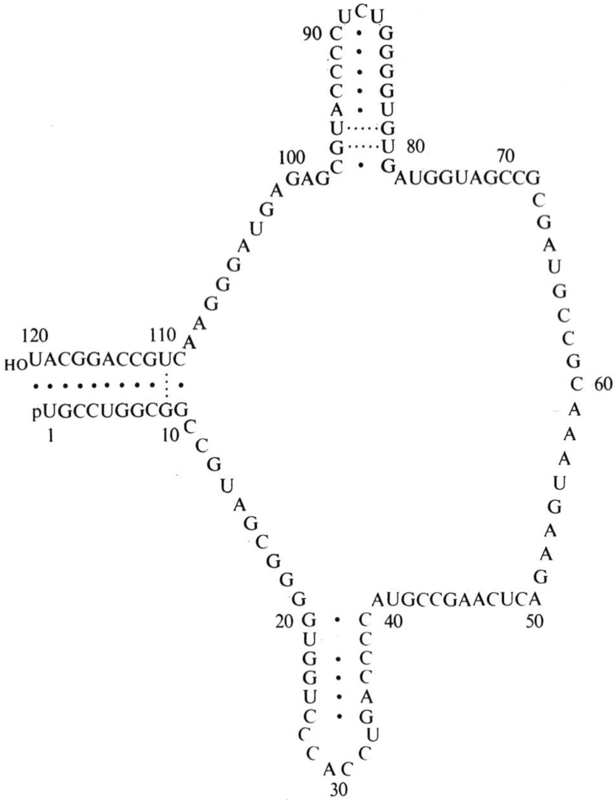

Fig. 12.18a. 5S RNA from *E. coli* (Brownlee *et al.*, 1967).

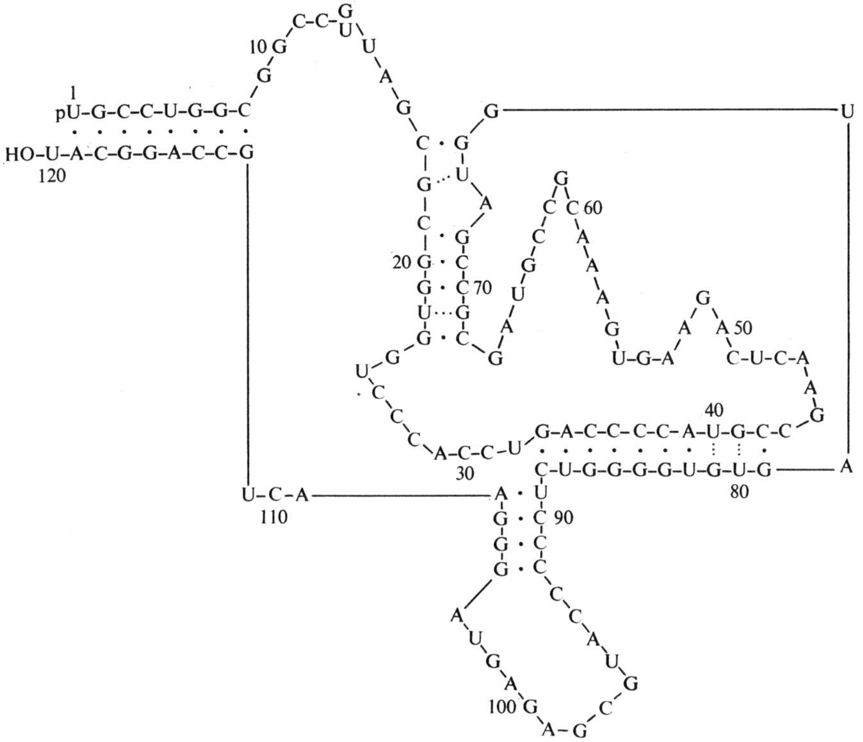

Fig. 12.18b. 5S RNA from *E. coli* (Tinoco, 1970).

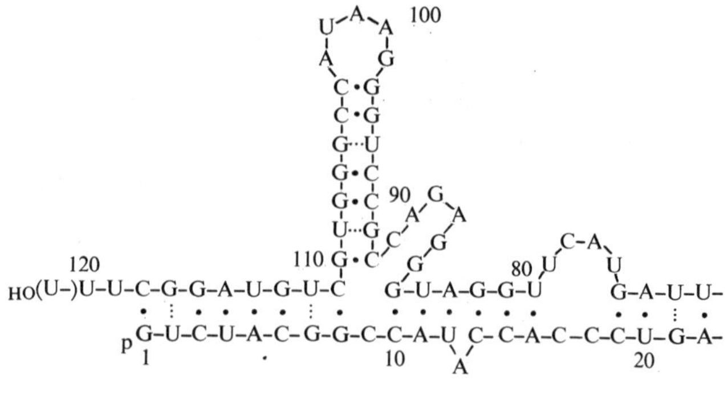

Fig. 12.19. 5S RNA from KB carcinoma cell (Forget and Weissman, 1970).

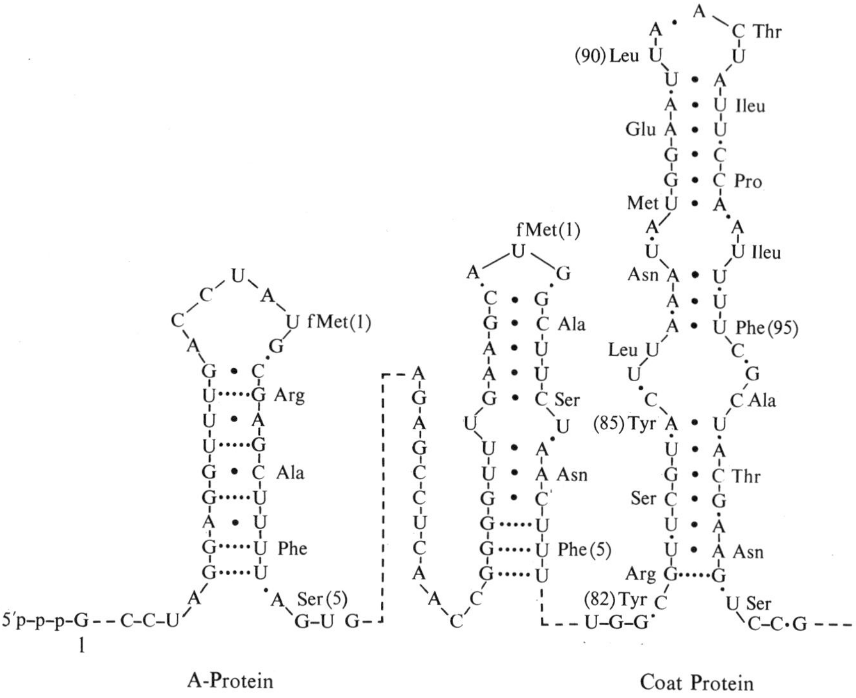

Fig. 12.20. R17 mRNA.

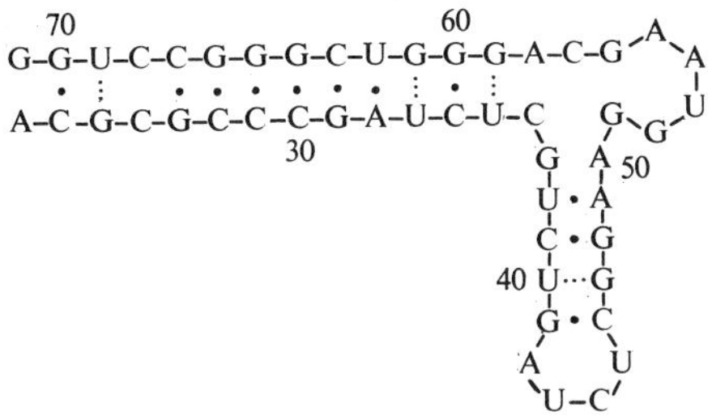
70
60
G–G–U–C–C–G–G–G–C–U–G–G–G–A–C–G
A–C–G–C–G–C–C–C–G–A–U–C–U–C
30
50
40

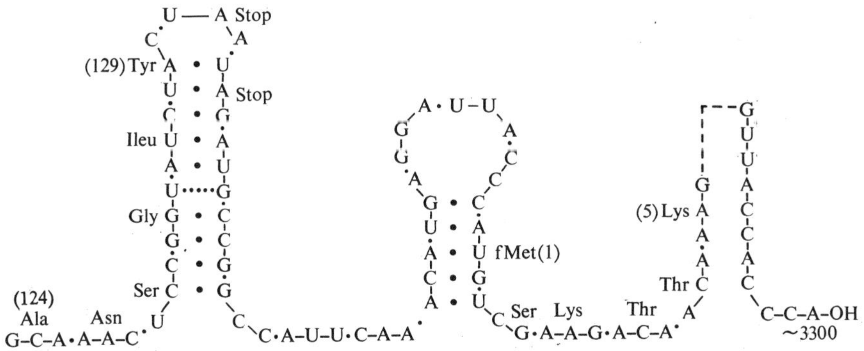
Stop
(129) Tyr
Stop
Ileu
Gly
Ser
(124)
Ala
Asn
G–C–A•A–A–C
C•A–U–U•C–A–A
fMet(1)
Ser
Lys
Thr
G•A–A–G•A–C–A
(5) Lys
Thr
C–C–A–OH
~3300

Synthetase

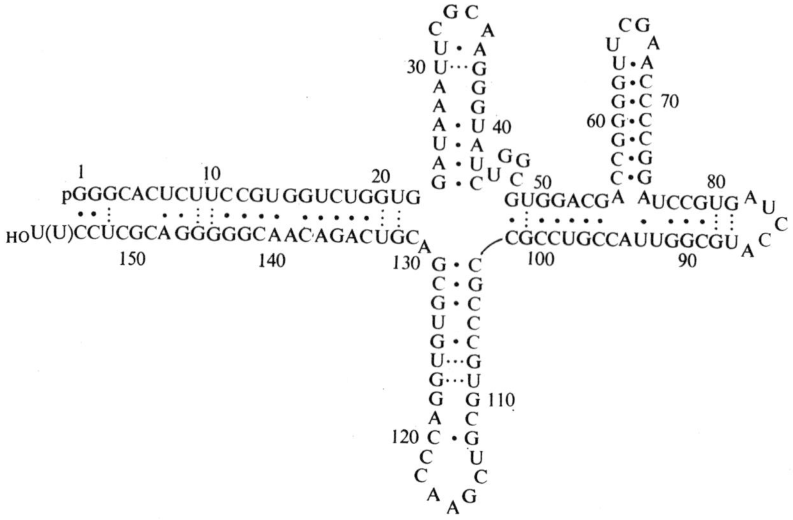

Fig. 12.21. VA RNA from adenovirus (Ohe and Weissman, 1970).

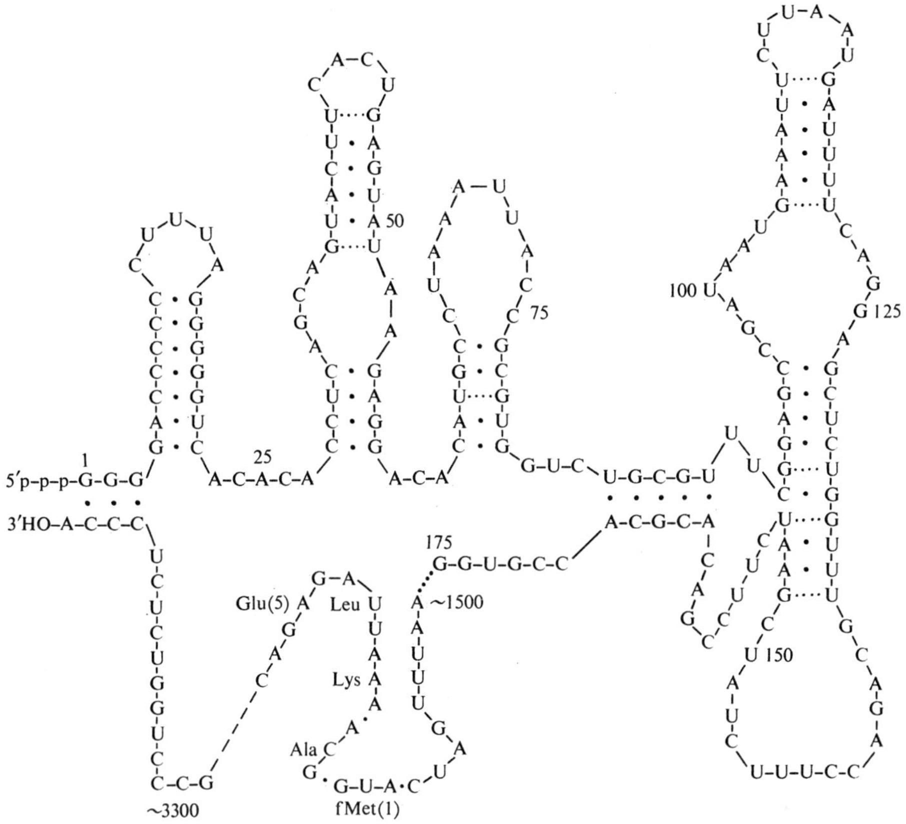

Fig. 12.22. Qβ mRNA.

REFERENCES

Baev, A. A., Venkstern, T. V., Mirzabekov, A. D., Krutilina, A. I., Li, L., and Axelrod, V. D. (1967). *Mol. Biol.*, **1**, 754.

Barrell, B. G., and Sanger, F. (1969). *FEBS Letters*, **3**, 275.

Brownlee, G. G., Sanger, F., and Barrell, B. G. (1967). *Nature*, **215**, 735.

Cory, S., Marcker, K. A., Dube, S. K., and Clark, B. F. C. (1968). *Nature*, **220**, 1039.

Dube, S. K., Marcker, K. A., Clark, B. F. C., and Cory, S. (1968). *Nature*, **218**, 232.

Dudock, B. S., Katz, G., Taylor, E. K., and Holley, R. W. (1969). *Proc. Natl. Acad. Sci.*, **62**, 941.

Forget, B. G., and Weissman, S. M. (1969). *J. Biol. Chem.*, **244**, 3148.

Goodman, H. M., Abelson, J., Landy, A., Brenner, S., and Smith, J. D. (1968). *Nature*, **217**, 1019.

Harada, F., Kimura, F., and Nishimura, S. (1969). *Biochim. Biophys. Acta*, **195**, 590.

Hashimoto, S., Miyazaki, M., and Takemura, S. (1969). *J. Biochem.* (Japan), **65**, 659.

Holley, R. W., Everett, G. A., Madison, J. T., and Zamir, A. (1965). *J. Biol. Chem.*, **240**, 2122.

Madison, J. T., Everett, G. A., and Kung, H. (1966). *Science*, **153**, 531.

Mizutani, T., Miyazaki, M., and Takemura, S. (1968). *J. Biochem.* (Japan), **64**, 839.

Ohe, K., and Weissman, S. M. (1970). *Science*, **167**, 879.

RajBhandary, U. L., and Chang, S. H. (1968). *J. Biol. Chem.*, **243**, 592.

Staehelin, M., Rogg, H., Baguley, B. C., Gensberg, T., and Wehrli, W. (1968) *Nature*, **219**, 1363.

Takemura, S., Murakami, M., and Miyazaki, M. (1969). *J. Biochem.* (Japan), **65**, 553.

Tinoco, I., Jr. (1970). Personal communication.

Yaniv, M., and Barrell, B. G. (1969). *Nature*, **222**, 278.

Zachau, H. G., Dutting, D., and Feldmann, H. (1966). *Z. Physiol. Chemie*, **347**, 212.

INDEX